VOLUME FIFTY FIVE

ADVANCES IN INSECT PHYSIOLOGY

Crop Protection

VOLUME FIFTY FIVE

ADVANCES IN INSECT PHYSIOLOGY

Crop Protection

Edited by

GUY SMAGGHE
Department of Plants and Crops,
Faculty of Bioscience Engineering,
Ghent University, Ghent, Belgium

Series Editor

RUSSELL JURENKA
Department of Entomology,
Iowa State University, Iowa, United States

ELSEVIER

ACADEMIC PRESS
An imprint of Elsevier

Academic Press is an imprint of Elsevier
125 London Wall, London, EC2Y 5AS, United Kingdom
The Boulevard, Langford Lane, Kidlington, Oxford OX5 1GB, United Kingdom
525 B Street, Suite 1650, San Diego, CA 92101, United States
50 Hampshire Street, 5th Floor, Cambridge, MA 02139, United States

First edition 2018

ISBN: 978-0-12-815079-5
ISSN: 0065-2806

For information on all Academic Press publications
visit our website at https://www.elsevier.com/books-and-journals

Publisher: Zoe Kruze
Acquisition Editor: Sam Mahfoudh
Editorial Project Manager: Peter Llewellyn
Production Project Manager: Abdulla Sait
Cover Designer: Miles Hitchen

Typeset by SPi Global, India

CONTENTS

CONTRIBUTORS

Sassan Asgari
Australian Infectious Diseases Research Centre, School of Biological Sciences, The University of Queensland, Brisbane, QLD, Australia

Olivier Christiaens
Department of Plants and Crops, Faculty of Bioscience Engineering, Ghent University, Ghent, Belgium

Feng Fang
Institute of Plant Protection, Shandong Academy of Agricultural Sciences, Jinan, PR China

Yulin Gao
State Key Laboratory for Biology of Plant Disease and Insect Pests, Institute of Plant Protection, Chinese Academy of Agricultural Sciences, Beijing, PR China

Wayne B. Hunter
USDA-ARS, Fort Pierce; AgTec, LLC, Port Saint Lucie, FL, United States

Vassiliki Labropoulou
Institute of Biosciences and Applications, NCSR "Demokritos", Athens, Greece

Zhongren Lei
State Key Laboratory for Biology of Plant Disease and Insect Pests, Institute of Plant Protection, Chinese Academy of Agricultural Sciences, Beijing, PR China

Donggang Li
Dezhou Academy of Agricultural Sciences, Dezhou, PR China

Ivan Meeus
Department of Plants and Crops, Faculty of Bioscience Engineering, Ghent University, Ghent, Belgium

Jinzhi Niu
Key Laboratory of Entomology and Pest Control Engineering, College of Plant Protection; Academy of Agricultural Sciences, Southwest University, Chongqing, China

Xiomara H. Sinisterra-Hunter
USDA-ARS, Fort Pierce; AgTec, LLC, Port Saint Lucie, FL, United States

Guy Smagghe
Department of Plants and Crops, Faculty of Bioscience Engineering, Ghent University, Ghent, Belgium

Jingchen Sun
College of Animal Science, South China Agricultural University, Guangzhou, People's Republic of China

Luc Swevers
Institute of Biosciences and Applications, NCSR "Demokritos", Athens, Greece

Liangde Tang
Environment and Plant Protection Institute, Chinese Academy of Tropical Agricultural Sciences, Haikou, PR China

Clauvis Nji Tizi Taning
Department of Plants and Crops, Faculty of Bioscience Engineering, Ghent University, Ghent, Belgium

Jin-Jun Wang
Key Laboratory of Entomology and Pest Control Engineering, College of Plant Protection; Academy of Agricultural Sciences, Southwest University, Chongqing, China

Shengyong Wu
State Key Laboratory for Biology of Plant Disease and Insect Pests, Institute of Plant Protection, Chinese Academy of Agricultural Sciences, Beijing, PR China

Xin Yuan
Agricultural Experimental Station, Zhejiang University, Hangzhou, PR China

Fan Zhang
Key Laboratory of Animal Resistance Research, College of Life Science, Shandong Normal University, Jinan, The People's Republic of China

Xian Cui Zhang
Key Laboratory of Animal Resistance Research, College of Life Science, Shandong Normal University, Jinan, The People's Republic of China

Yongchao Zhao
Institute of Biosciences and Applications, NCSR "Demokritos", Athens, Greece; College of Animal Science, South China Agricultural University, Guangzhou, People's Republic of China

PREFACE

Insects! They are with an estimated 10 quintillion individuals on Earth and maintain a position of ecological pre-eminence for over 400 million years; they have witnessed the rise and fall of dinosaurs; they have survived at least four major cataclysms that resulted in planet-wide extinctions; and they continue to thrive despite mankind's best efforts at eradication. Apart from important ecological and economic services such as pollination and biological control, they also cause problems by damaging crops and food production in agriculture, parasitizing livestock, or being a nuisance and health hazard to humans and animals. Although dollar values for these losses are nearly impossible to calculate, especially when they affect human health and welfare, economists generally agree that insects consume or destroy around 10% of the gross national product in large, industrialized nations and up to 25% of the gross national product in some developing countries.

"The good, the bad, and the ugly in microbe world! Sounds like a western movie!" In the recent years, evidence has accumulated that microbes directly or indirectly influence insect interactions at individual, population and community levels. Microbes are very diverse and represent all the great kingdoms of life; the major groups are viruses, bacteria, algae, fungi and protozoa.

Therefore, this issue of *AIP* contains a collection of seven chapters, covering a range of biological systems and recent experimental advancements. It wishes to gather knowledge and increase our understanding on microbe-mediated effects in pest insects and also beneficials as pollinating bees. Indeed it is an emerging field of research with considerable implications for sustainable agriculture, protection of crops and so safeguarding food production to feed an increasing world population of nearby 9 billion.

Today, it is clear, of course, that the modern NGS technologies provide goldmines of information, increasing enormously our knowledge of the biology of the insects and their immunity system. Also the postgenomic Nobel Prize-winning RNAi technology revolutionized the field to understand gene functions and unique applications in pest insect control so came possible. Chapter 1 by Niu et al. rethinks on the use of RNAi to control important pest insect and provides also new ideas on the synergy between the siRNA, as the natural viral immune system, and the miRNA pathway. Chapter 6 by Hunter and Sinisterra-Hunter further demonstrates how to protect citrus trees by a nontransgenic plant RNAi approach via the impregnation of trees with dsRNA against essential genes of the Asian citrus psyllids that are the vectors of the feared greening disease bacteria. This success is a big step forward in the field. However, Chapter 2 by Asgari focuses on the latest, more fundamental achievements with microRNAs as regulators of insect host–pathogen interactions and insect immunity. A fascinating emerging direction of research is the development of biotechnology platform for high-throughput screening. Chapter 5 by Zhao et al. is looking beyond baculoviruses, providing a new additional platform based on insect RNA viruses. We hope these chapters will spur interest and enthusiasm for the field of microbial mediation of insect immune interactions and stimulate researchers to take up these current and future challenges in this field of pest control.

Another important challenge is that in the real world microbes function in a whole community. In this context, it can be said that microbes are everywhere, but we can't see them. They make us sick, but we can't live without them. Indeed the strong impacts of the bacteria living inside the insect body have spurred interest in how to manipulate them to increase the success of beneficial insects or reduce the performance of pest insects, another big challenge in the field of sustainable agriculture. One of the obstacles for efficient crop pollination is a good bee health, and Chapter 3 by Meeus and Smagghe dwells deep on the implications of the eusociality behaviour in the honeybee queen on parasite-mediated competition. However, Chapter 4 by Zhang and Zhang sheds light on the presence and community context of the microbes in the gut of cockroaches and how the interaction is with entomopathogenic fungi to biocontrol these important pest insects that live in a wide range of habitats, as human households, around the world. Finally, Chapter 7 by Wu et al. stimulates further the discoveries on the development of new microbial control agents against the sucking pest insects of thrips, with insecticide-resistant populations that cause damage in many crops worldwide. Yes these continuous discoveries on the diversity and roles of microbial communities are stimulating the development of new pest control strategies as entomopathogenic microbes.

Guy Smagghe

CHAPTER ONE

Rethink RNAi in Insect Pest Control: Challenges and Perspectives

Jinzhi Niu*^,†^, Clauvis Nji Tizi Taning‡, Olivier Christiaens‡, Guy Smagghe‡, Jin-Jun Wang*^,†^

*Key Laboratory of Entomology and Pest Control Engineering, College of Plant Protection, Southwest University, Chongqing, China
†Academy of Agricultural Sciences, Southwest University, Chongqing, China
‡Department of Plants and Crops, Faculty of Bioscience Engineering, Ghent University, Ghent, Belgium

Contents

Abstract

RNA interference (RNAi), usually referring to the small interfering (si)RNA pathway, is promising in the development of a new generation of insect pest control products. However, prior to its exploitation for insect pest control, it is important to consider potential limiting factors, such as immune response and fitness cost, RNAi efficiency and dsRNA degradation, and virus-encoded suppressor of RNAi factors in the development of the RNAi-based pest control strategy. Additional challenges such as lack of feasible dsRNA delivery methods in practice, low efficiency in pest control capacity and evolution of resistance to RNAi have largely constrained the application of RNAi in

Advances in Insect Physiology, Volume 55
ISSN 0065-2806
https://doi.org/10.1016/bs.aiip.2018.07.003

practice. This review highlights how insect ecology and integrated pest management principles, taking into account RNAi, could be exploited to promote sustainable insect pest control.

1. RNAi IN INSECT PEST CONTROL

Insect pests cause huge crop losses globally and the use of broad-spectrum chemical pesticides to control them is also responsible for environmental contaminations as well as the threat to pollinators (Goulson et al., 2015; Guedes et al., 2016; Rundlöf et al., 2015). Moreover, the development of resistance to these chemical pesticides also challenges their sustainable use as pest control agents (Guedes et al., 2016; Van Leeuwen and Dermauw, 2016). Therefore, novel sustainable and environmentally friendly approaches that could complement broad-spectrum chemical pesticides in the management of insect pests are urgently needed (Douglas, 2017), and one of such approaches could be RNA interference (RNAi).

RNAi is a conserved mechanism for posttranscriptional sequence-specific gene silencing and is initiated by the introduction of double-stranded RNA (dsRNA) (Baulcombe, 2004; Joga et al., 2016; Zotti et al., 2018). RNAi can be sorted into three pathways depending on the type of small RNA biogenesis and the Argonaute protein involved: microRNA (miRNA), small interfering RNA (siRNA) and piwi-interacting RNA pathways. Among them, the siRNA pathway has shown the most potential in insect pest control. Briefly, when exogenous dsRNA designed to target an endogenous gene is delivered to the target pest, it triggers the slicing activity of Dicer, resulting in the long dsRNA being processed into short 21–22 nt siRNAs. These siRNAs are then incorporated into a complex together with Argonaute and other proteins, forming the RNA-induced silencing complex. In this complex, one of the small RNA strands is selected as the guide strand and serves as a sequence-specific guide to cleave the target endogenous mRNA through sequence-specific complementary binding (Kim et al., 2009). By targeting the expression of a vital endogenous gene necessary for survival, RNAi can be exploited to induce target gene silencing and pest mortality, demonstrating its potential as a promising tool in insect pest management (Joga et al., 2016; Zotti et al., 2018).

Prior to the exploitation of RNAi in pest management, several important aspects about the nature of RNAi in insects need to be understood: (i) RNAi (siRNA) is one of the key antiviral pathways in insects

(Gammon and Mello, 2015; Marques and Imler, 2016; Swevers et al., 2018). Although insects are known to lack an adaptive immunity system, there are crosstalks between RNAi and other non-RNAi immune mechanisms, reflecting the sophisticated immunity network of insects (Gammon and Mello, 2015). (ii) The fitness costs of RNAi in antiviral immunity. The activity of the RNAi pathway like other immune pathways could be significantly induced upon viral infections (Niu et al., 2014), which could cause a disturbance in the existing homeostasis, resulting in a trade-off with other individual parameters (Schwenke et al., 2016). (iii) RNAi-based pest control relies on the silencing of crucial genes for survival in the target insect pest, ultimately causing mortality and/or inhibition of the pest populations. However, the potential of RNAi (siRNA) in inducing gene silencing may be limited within certain thresholds, resulting in poor gene silencing efficiency (Christiaens et al., 2018a; Niu et al., 2018). (iv) Direct degradation of dsRNA in the insect gut as well as in the hemolymph by nucleases can decrease gene silencing efficiency (Christiaens et al., 2014; Guan et al., 2018; Liu et al., 2012; Wang et al., 2016). (v) Virus-encoded suppressors of RNAi factors can also inhibit the activity of the RNAi pathway (Gammon and Mello, 2015; Kolliopoulou et al., 2017; Swevers et al., 2013). The increased use of next-generation sequencing has led to the detection of novel viruses, especially RNA viruses, among which some have been reported to closely interact with some components of the RNAi machinery (Remnant et al., 2017; Webster et al., 2015). Further studies will be required to better understand the detail role of RNAi as a defence system against these new viruses and how these viruses counter the RNAi mechanism.

2. CHALLENGES IN APPLYING RNAi FOR PEST CONTROL

Until now, the approach of using RNAi in insect pest control has only been reported for commercialization in the western corn rootworm (*Diabrotica virgifera virgifera*) (Fishilevich et al., 2016; Palli, 2014). While most RNAi-based insect pest control studies are still in the lab stages (Joga et al., 2016; Zotti et al., 2018), others are only meaningful for functional gene analyses, far away from pest control in practice. The reasons for this are variable and could be summarized in some general factors/points as discussed below.

2.1 dsRNA Delivery in Practice

The overall success of using RNAi for pest control is dependent on the mode of delivery of dsRNA. Since gene silencing is only limited to cells that take up the target gene dsRNAs, dsRNA delivery remains a major consideration in the development of an RNAi-based insecticide (Niu et al., 2018; Yu et al., 2016; Zotti et al., 2018). Different delivery systems have been studied in various insect species and the main dsRNA delivery methods tested so far include soaking, feeding, injection and transgenic plants expressing dsRNA.

dsRNA uptake in insects could be divided into two levels: (i) dsRNA uptake from the environment into the insect body. This could either be from the ingestion of dsRNA from diet or/and direct uptake via the insect skeleton (exocuticle). Current practical methods of dsRNA delivery include either transgenic plants modified to express insect pest-specific dsRNAs (for example in *D. virgifera virgifera*) or sprayable dsRNA formations (for example in *Leptinotarsa decemlineata*) (Fishilevich et al., 2016; Palli, 2014). The transgenic approach of dsRNA delivery offers the best potential for practical field application and has been widely studied in various key insect pests and spider mites (Christiaens et al., 2018a; Niu et al., 2018). Alternatively, with the development of microbial production systems for exogenous dsRNAs, pest-specific dsRNAs can now be produced in large scales and at affordable prices for direct application in the field, by spraying (Zotti et al., 2018). (ii) Cellular uptake and the stability of dsRNA in insects after uptake of the dsRNA from the environment into the insect body. Limiting factors, such as direct degradation (Guan et al., 2018; Wang et al., 2016) and entrapment of internalized dsRNA in endosomes (Yoon et al., 2017), have recently been shown to be crucial in achieving a successful RNAi during this step of dsRNA delivery.

2.2 Efficiency of RNAi in Pest Control

The success of RNAi in pest control depends on its efficiency in causing mortality to the targeted pests, ultimately leading to a decrease in the insect pest population under the economic threshold. The choice of a target gene that could lead to such a level of pest control is based on: genotype—the relative gene silencing level of the targeted gene via RNAi; phenotype—the function of the targeted gene, which implies that the role of this gene must be crucial for the survival of the insect. However, gene silencing efficiency is not only influenced by the target gene but also by the method of dsRNA delivery as mentioned earlier. Other associated factors, based on the nature of the target gene that could affect gene silencing efficiency include: the absolute expression level of the target gene, the relative expression level of the target gene at various time points, tissue-specific expression pattern of the target gene as well as the expression of paralogous genes with similar functions (Christiaens et al., 2018a; Cooper et al., 2018).

2.3 Resistance Development to RNAi

As observed with most insecticides, resistance to RNAi-based insecticides is bound to happen. Therefore, the key questions are what potential mechanisms could be involved in resistance development against RNAi-based insecticides and which possible sustainable control strategies could be explored/developed to delay/manage these resistance challenges. In a recent study, the first case of field-resistant strains of *D. virgifera* against dsRNA (dsDvSnf7) was reported (Khajuria et al., 2018). A 130-fold resistance was observed in the strain feeding on transgenic plants expressing dsDvSnf7 in comparison with the susceptible strain reared on nontransgenic plants. In addition, the resistant strain was also tolerant to other dsRNA treatments, which is suggesting that the RNAi-based resistance observed in *D. virgifera* is not related to the target gene, but probably to the general dsRNA uptake mechanism. Although the mechanism preventing dsRNA uptake in the resistant strain is still unknown, various factors involved in dsRNA uptake both from the environment and in the insect body could be involved. In a study on migratory locust, it was reported that inter- and intrapopulation variations could affect RNAi efficiency (Sugahara et al., 2017). This implies that variations of RNAi efficiency in various populations could also represent a driver of tolerance/resistance against RNAi-based insecticides.

3. RNAi IN THE CONTEXT OF INSECT ECOLOGY

From an ecological point of view, insect pests in agriculture, being herbivores, are in constant interaction with plants and pathogens which results in coevolution. During feeding, insect pests trigger the defence response of plants which include the genes encoding defensive proteins and enzymes for the synthesis of toxic metabolites. On the other hand, the insect pests evolved to interfere with these plant defences, by either inhibiting the plant defence mechanism or adapting to the detrimental effects (Stahl et al., 2018). Another important interspecific relation is the interaction between insects and pathogens. The infection of insects by pathogens triggers the immunity of the insect hosts, which includes microbial recognition and subsequent immune signal transduction (Buchon et al., 2014). RNAi technology is a manipulator of gene expressions, implying that the key genes in insect pests, required for maintaining homeostasis with the interacting host plants and pathogens, could be targeted, presenting an alternative approach in RNAi-based pest control strategy (Eakteiman et al., 2018; Yu et al., 2016; Zhang et al., 2017b). For instance, silencing of genes involved in feeding in aphids, such as effectors to plant defences, resulted in significant mortality (Yu et al., 2016). Similarly, targeting the detoxification gene, glutathione *S*-transferase gene, by phloem-mediated RNAi in the whitefly, *Bemisia tabaci*, showed long-term inhibition of the population (Eakteiman et al., 2018). Moreover, by depleting insect immunity genes, such as toll receptors, the virulence of pathogens can be increased (Zhang et al., 2017b).

Thus, the concept of combining RNAi with plant resistance or/and pathogens termed as RNAi^{plus} (Niu et al., 2018), could be helpful in solving current challenges of RNAi in insect pest control (Fig. 1): (i) RNAi could be used to assist pest control strategies such as plant-produced toxins, insecticides, pathogens, etc. (ii) Neither RNAi nor plant-produced toxins, insecticides and pathogens are fully efficient in controlling pests, however, the combination could lead to efficient pest control. (iii) The combination of RNAi and other pest control strategies could largely delay the development of resistance. For instance, variable combinations could be used, taking into account synergistic effects of various pathogens or various insecticides or plant toxins, with the RNAi-based control strategy. In such scenario, the RNAi-based strategy will be designed to target various genes involved in maintaining homeostasis between the plant host, insect pests and insect pathogen, thereby following an ecological principle rather than targeting a single factor which can easily pressure the insects to develop resistance. In addition, with environmental RNAi or/and cross-kingdom RNAi, the delivery of RNAi could also be enhanced either through the host plants or the insect pathogens (Cai et al., 2018; Wang et al., 2017a).

4. dsRNA DELIVERY

The efficient delivery of dsRNA is crucial in moving RNAi-based insect pest control from the lab to the field. From an ecological viewpoint, dsRNAs not only move within an organism but they can also transfer from the environment to the organism (environmental uptake) and between interacting organisms (cross-kingdom dsRNA trafficking, thereby subsequently inducing gene silencing in the targeted organism; Cai et al., 2018; Wang et al., 2017a).

4.1 Environmental RNAi and Insect Pest Control

The environmental RNAi approach can be exploited in the application of sprayable dsRNA formulations for insect pest control, similar to the application of classical chemical insecticides. The delivery and uptake of insecticides in insect pests mainly happen through the insect skeleton (exocuticle) and/or ingestion during feeding on host plants. In the potential RNAi-mediated control of chewing pests, the spray-dsRNA formations can be successful and such applications, for example against *L. decemlineata*, are on their way to be commercialized (Palli, 2014). However, sucking-piercing pests, such as aphids, whiteflies as well as spider mites, which are either feeding on the plant phloem or cells, require the dsRNA uptake via the insect integument directly and/or via ingestion of the dsRNA from plant phloem or cells. For sprayable dsRNA formulations, this presents an extra challenge. Using a topical delivery approach, significant gene silencing of five *CYP4* genes was achieved and resulted in suppressed insecticide resistance in *Diaphorina citri* (Killiny et al., 2014). A soaking-based dsRNA delivery was also used successfully in targeting genes in piercing–sucking pests (Luan et al., 2013; Niu et al., 2018). If dsRNA could be taken up by plant leaves then these sucking pests could be targeted via ingestion of dsRNA during feeding on the plant phloem or cells (Wang and Jin, 2017). Recent research demonstrated that mite genes could be silenced through feeding on leaves presprayed with dsRNA, even though the gene silencing efficiencies were still limited (Niu et al., 2018). However, for the phloem-sucking insects, dsRNAs are mainly delivered through the stem/leaf in the lab (Luan et al., 2013; Yu et al., 2016), which are not an applicable delivery strategy in the field. Therefore, studies elucidating the mechanism of dsRNA uptake by insect cuticle or plant leaf are required in the development of dsRNA-sprayable pest control strategies. Two aspects of fundamental studies are currently needed: (1) the mechanism of dsRNA uptake by insects or plant leaves, which can be achieved by using labelled dsRNA/siRNA to visualize the mobility of dsRNA/siRNA and (2) exploring the ingredients or methods that could enhance the stability and the absorbing efficiency of dsRNA by insects and plants. For example, adjuvants used for chemical pesticides can be evaluated for their ability to improve the uptake of dsRNA by the insect/plant leaf. Also, liposomes, polymers and peptides have been successfully used to increase persistence and cellular uptake of dsRNA in insects (Christiaens et al., 2018b; Gillet et al., 2017; Taning et al., 2016). Nanoparticles, such as nanosheets, have potential for applications of dsRNA by topical delivery, thus it may also be useful in delivery insect pest-specific dsRNAs (Mitter et al., 2017).

4.2 Cross-Kingdom RNAi and Insect Pest Control

Based on the concept of cross-kingdom RNAi, plants and microbes can be exploited to mediate dsRNA production and delivery. Transgenic plants expressing dsRNA have been shown to be useful in insect pests control (Zhang et al., 2017a). In the pioneering work for the control of *D. virgifera virgifera*, *Snf7*-dsRNA was transcribed and delivered via transgenic corn (Fishilevich et al., 2016). The engineered transgenic plants expressing sequence-specific dsRNA to target pests are currently investigated in various organisms (Zhang et al., 2017a). Several studies have found that dsRNA produced by nuclear gene expression in these transgenic plants, is being processed by the plant's own RNAi machinery, resulting in mainly small RNAs being available for the insects (Thakur et al., 2014). This could potentially affect the efficacy of these host plants in achieving a high degree of gene silencing in pest insects. Therefore, the expression of long dsRNAs in plastids where the RNAi machinery is absent, offers a particularly promising strategy that could lead to a more efficient silencing and therefore a more efficient management of the pest species (Jin et al., 2015; Zhang et al., 2015, 2017a). Still, transgenic plant-based RNAi strategies in insect pest control face some limitations, such as transformation ability of many crops, genetic stability of a crop species as well as public concerns about genetically modified organisms. In addition, crops in the field could be faced with a wide range of pest species and a fluctuating dynamic of the populations of pests in various seasons/years. Some insect pests mainly infest apparent hosts (long-lived, mainly woody plants), making it difficult to use this strategy in the control of such pests since there is less flexibility in changing to another transgenic variety for example when confronted with new or additional pest species (Fig. 1).

Microbes, another important ecological interacting factor of insect pests, also provide an alternative way for dsRNA production and delivery. The bacterial model *Escherichia coli* is not only used for large-scale production of dsRNA but also the ingestion of engineered *E. coli* in expressing dsRNA targeting *Chitinase* directly, increased the mortality and reduced the weight of the feeding larvae of *Mythimna separata* (Ganbaatar et al., 2017). In addition, the engineered fungi *Saccharomyces cerevisiae*, expressing specific dsRNA targeting *Drosophila suzukii*, led to a significant decrease of fitness of the targeted pest (Abrieux and Chiu, 2016). Besides these model microbes, an ecological design focused on exploiting the insect symbiotic bacteria for trauma-free sustainable delivery of insect pest-specific

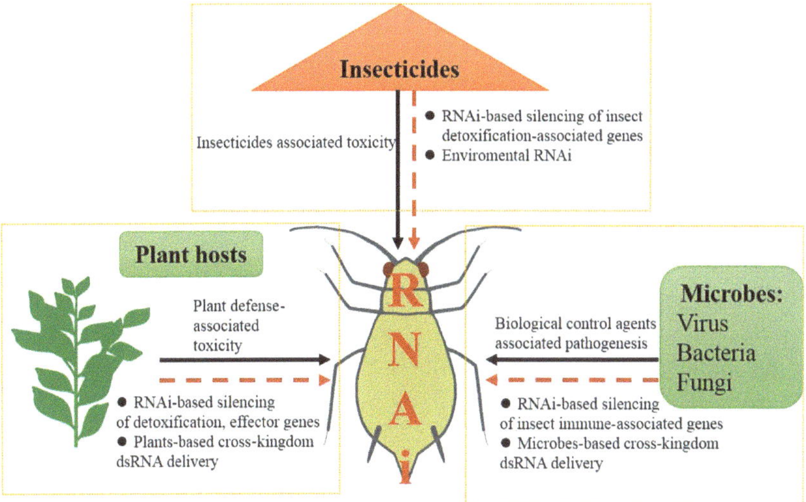

Fig. 1 The illustration of RNAi^{plus}-based dsRNA delivery and integrated with other pest control strategies.

dsRNA (Miranda and Paul, 2017). Recent studies also identified a great number of insect viruses, especially RNA viruses (Shi et al., 2016). Based on plasmid-based reverse genetics systems, engineered viruses could also produce host-specific dsRNA, thus providing potential in pest control. Viruses are obligated to use the host nucleic acid and protein synthesis machinery and dsRNA is produced as an intermediate molecule during RNA virus replication, while RNA hairpins are present after transcription with DNA viruses (Kolliopoulou et al., 2017). A first proof of concept was recently provided by Taning et al. (2018) who engineered a Flock House Virus (FHV) to produce *D. melanogaster* specific dsRNA and managed to cause a high degree of RNAi silencing upon viral infection, leading to a strong mortality, compared to the wild-type FHV-infected flies.

Studying the nature of cross-kingdom dsRNA movement is also useful in obtaining an optimal pest control strategy. For instance, *Varroa* mites are one of the major drivers in bee decline (Goulson et al., 2015), and the control of this pest is crucial to protect bees from mites transmitted bee viruses. Intriguingly, dsRNA targeting mite genes can be transferred to mites through feeding of dsRNA-supplemented syrup to bees (Garbian et al., 2012). In addition, the dsRNA can also be horizontally transferred between individuals and also across generations in bees (Maori et al., 2018).

5. SYNERGY

Chemical insecticides are still the most efficient and easy-to-adopt methods for insect pest control, even though the concept of integrated pest management (IPM) strategy has been widely accepted and incorporated in crop protection in the efforts to optimize a sustainable pest control. Current IPM programs are still considerably less efficient in comparison with the sum of single-used crop protection strategies. Therefore, a practical approach synergistically combining traditional and novel approaches is currently needed (Douglas, 2017; Stenberg, 2017). RNAi technology could be one of these novel approaches and could be considered an "indirect" pest control method, in contrast to insecticides, microbes, plant-related toxins as well as *Bacillus thuringiensis* (*Bt*)-related toxins which are directly harmful to insect pests. Hence, the question is to what extend we can explore this mechanism in pest control? A recent study showed that despite the ability to choose and alter the target genes, RNAi technology is not immune to resistance build-up in pest insect populations (Khajuria et al., 2018). The direct control methods, such as insecticides, microbes, plant-related toxins as well as *Bt*-related toxins, have also been constrained by resistance after long-term application. The opportunity of combining different control methods will be crucial in avoiding rapid occurrence of resistance against these pest control methods. Several strategies, including classical insecticides and/or pathogens (including pathogen-derived pesticides such as *Bt* toxins) have been proposed as being good partners to RNAi in an IPM context (Fig. 1).

5.1 Management of Current Insecticide and *Bt* Resistance

5.1.1 Insecticide Resistance

RNAi has been successfully used to investigate insecticide target genes, genes involved in insecticide detoxification and resistance, or nuclear receptors involved in the regulation of detoxification genes (Heidel-Fischer and Vogel, 2015; Kim et al., 2015). Silencing of *acetylcholinesterase*, *cytochrome P450 monooxygenases*, *carboxyl/cholinesterases*, *glutathione S-transferases*, *ATP-binding cassette transporters*, etc., that are associated with insecticide detoxification and/or resistance can increase the sensitivity of the targeted pests to insecticides, thereby reducing the level of resistance and pesticide applications (Clements et al., 2017; Kim et al., 2015; Kishk et al., 2017; Wang et al., 2017b; Xu et al., 2017). In the future, combining RNAi with insecticide needs to be evaluated in the manner of application level, especially looking for potential synergistic effects between the targeted insecticide detoxification and/or resistance gene and the toxicity of the relative insecticides.

5.1.2 **Bt** *Resistance*

Evolution of resistance to *Bt* Cry toxins has been genetically linked to mutations or expression alterations of receptor genes such as *cadherin*, *aminopeptidase-N*, *alkaline phosphatase* and *ATP-binding cassette transporter subfamily C*. Direct depletion of these receptor genes results in resistance to *Bt* Cry toxins, while RNAi targeting the MAPK upstream gene, which regulates the expression level of *Bt* receptor genes, can restore susceptibility to *Bt* in resistant strains (Guo et al., 2015; Xiao et al., 2016). Using RNAi technology targeting *Bt* resistance-associated regulation genes in combination with *Bt*-based products could be a next step, trying to counter the resistance, but it requires further exploration for both transgenic plant approaches and *Bt* Cry toxin-based biopesticides (Guan et al., 2017; Ni et al., 2017).

5.2 Enhancement of Insect Pest Pathogens Virulence

Incorporating entomopathogens, such as viruses, bacteria and fungi, as biological control agents in arthropod pest control has been a long-time interest of crop protectors (Lacey et al., 2015). Coevolution of these microbial control agents and their hosts has resulted in an equilibrium where the host and pathogen coexist (Buchon et al., 2014). Inhibition of the innate immunity or the obstacle that constrains the pathogens infection could be a potential mode of biological pest control (Liu et al., 2017). Largely thanks to metagenomics, more and more microbes in insects have been discovered and keep being discovered, providing new candidates for further exploitation as microbial control agents. Using RNAi, knockdown of the *B. tabaci Toll* gene increased the mortality of a challenge with destruxin A (Zhang et al., 2017b). Furthermore, in locusts an increased virulence of a specific fungal pathogen *Metarhizium acridum* expressing dsRNAs targeting the host *ATPase subunit* genes was observed (Hu and Xia, 2018). Targeting genes involved in the immune responses of the soybean pod borer *Leguminivora glycinivorella* resulted in relatively high larval mortality rates (Ran et al., 2018). Currently, the studies in insect–pathogen interactions revealed a number of immune/pathogen defence-associated genes, while novel pathogens and host response genes are also intensively reported thanks to the applications of next-generation sequencing (Douglas, 2018). Further studies are required to investigate the synergy between RNAi-based gene silencing and virulence of pathogens to make progress in this field.

5.3 Enhancement of Plant Host Defence to Herbivore Pests

During feeding of herbivore pests, plants detect the attack and organize an efficient defence response, which involves defensive proteins for synthesis of toxic metabolites. At the same time, insects can produce effectors to inhibit detection or adapt to their detrimental effect, such as detoxification (Stahl et al., 2018). Breakdown of this balance between plants and herbivore pests through targeting effectors or detoxification genes by RNAi shows potential as a pest control strategy (Eakteiman et al., 2018; Niu et al., 2018; Yu et al., 2016). In addition, some of the detoxification genes associated with insecticides and plant-derived toxins overlap (Dermauw et al., 2018), which means that RNAi applications could potentially target both issues at once and achieve a synergistic effect.

6. CONCLUSIONS AND REMARKS

Now that it is clear that RNAi holds potential as a pest control strategy, more fundamental studies are needed to further develop specific field application methods. The mode of action of RNAi in killing pests is unique and distinct with that of classical insecticides, microbes, plant-related toxins as well as *Bt*-related toxins which are directly harmful to pests. RNAi needs to induce negative effects to the targeted pests by silencing essential genes, thus it depends on the level of exploitation of this insect's 'endogenous gene regulation mechanism'. The nature of RNAi in insects has mainly evolved as an antiviral defence mechanism, which is part of the innate immunity in insects. Therefore, with the exploitation of this mechanism in pest control, it is necessarily to consider its immune response and fitness cost. In addition, the challenges of RNAi-based pest control such as dsRNA delivery in practice, efficiency of its pest control capacity, and resistance problems, also constrain the application of this strategy. Insect herbivores are normally involved into two important aspects of coevolution, plants and microbes. Utilization of the synergy of RNAi with plants and microbes presents a potentially effective strategy in enhancing dsRNA delivery, efficiency, specificity, as well as reducing the risk in developing resistance to RNAi. Additionally, we recommend the following studies which would also be helpful in implementing a better RNAi-based insect pest control program: (i) both plants and microbes could be used in delivering dsRNA and the design of these dsRNA targets in enhancing the plant defence mechanisms or virulence of microbes still requires further evaluation in optimizing this strategy based on both the delivery and the synergy; (ii) aside from the siRNA pathway, miRNAs also play an important role, not only in internal gene regulation but also in cross-kingdom regulation with interacting organisms (Cai et al., 2018; Weiberg et al., 2015). The mobile miRNAs mechanisms have great potential in exploitation for crop protection together with RNAi (siRNA); (iii) nontarget effects to beneficial insects require careful evaluation. Issues such as the fate of dsRNA in the environment and potential adverse effects on nontarget organisms, including beneficial insects such as pollinators, predators and parasitoids, have to be further investigated.

ACKNOWLEDGEMENTS

The authors acknowledge support of the National Key R&D Program of China (2017YFD0200900) and the Foundation Project of Southwest University (SWU114049) in China, and the Research Foundation–Flanders (FWO) in Belgium. J.N. is a recipient of China Postdoctoral Science Foundation Grant (2018T110939) and Postdoctoral Special Research Funds of Chongqing Municipal (Xm2017017). O.C. is a recipient of a postdoctoral fellowship of the Research Foundation–Flanders (FWO), and C.N.T.T. is a recipient of a postdoctoral fellowship of the Special Research Fund of Ghent University (BOF-UGent).

Competing interests: The authors declare no competing financial interests.

REFERENCES

Abrieux, A., Chiu, J.C., 2016. Oral delivery of dsRNA by microbes: beyond pest control. Commun. Integr. Biol. 9, e1236163.

Baulcombe, D., 2004. RNA silencing in plants. Nature 431, 356–363.

Buchon, N., Silverman, N., Cherry, S., 2014. Immunity in *Drosophila melanogaster*—from microbial recognition to whole-organism physiology. Nat. Rev. Immunol. 14, 796.

Cai, Q., He, B., Kogel, K.-H., Jin, H., 2018. Cross-kingdom RNA trafficking and environmental RNAi—nature's blueprint for modern crop protection strategies. Curr. Opin. Microbiol. 46, 58–64.

Christiaens, O., Swevers, L., Smagghe, G., 2014. DsRNA degradation in the pea aphid (*Acyrthosiphon pisum*) associated with lack of response in RNAi feeding and injection assay. Peptides 53, 307–314.

Christiaens, O., Dzjambazova, T., Kostov, K., Arpaia, S., Joga, R.M., Urru, I., Sweet, J., Smagghe, G., 2018a. Literature review of baseline information on RNAi to support the environmental risk assessment of RNAi-based GM plants. EFSA 15, 1424E.

Christiaens, O., Tardajos, M.G., Martinez Reyna, Z.L., Dash, M., Dubruel, P., Smagghe, G., 2018b. Increased RNAi efficacy in *Spodoptera exigua* via the formulation of dsRNA with guanylated polymers. Front. Physiol. 9, 00316.

Clements, J., Schoville, S., Peterson, N., Huseth, A.S., Lan, Q., Groves, R.L., 2017. RNA interference of three up-regulated transcripts associated with insecticide resistance in an imidacloprid resistant population of *Leptinotarsa decemlineata*. Pestic. Biochem. Physiol. 135, 35–40.

Cooper, A., Silver, K., Jianzhen, Z., Park, Y., Zhu, K.Y., 2018. Molecular mechanisms influencing efficiency of RNA interference in insects. Pest Manag. Sci. https://doi.org/10.1002/ps.5126 (in press).
Dermauw, W., Pym, A., Bass, C., Van Leeuwen, T., Feyereisen, R., 2018. Does host plant adaptation lead to pesticide resistance in generalist herbivores? Curr. Opin. Insect Sci. 26, 25–33.
Douglas, A.E., 2017. Strategies for enhanced crop resistance to insect pests. Annu. Rev. Plant Biol. 69, 3.1–3.24.
Douglas, A.E., 2018. Omics and the metabolic function of insect–microbial symbioses. Curr. Opin. Insect Sci. 29, 1–6.
Eakteiman, G., Moses-Koch, R., Moshitzky, P., Mestre-Rincon, N., Vassão, D.G., Luck, K., Sertchook, R., Malka, O., Morin, S., 2018. Targeting detoxification genes by phloem-mediated RNAi: a new approach for controlling phloem-feeding insect pests. Insect Biochem. Mol. Biol. 100, 10–21.
Fishilevich, E., Vélez, A.M., Storer, N.P., Li, H., Bowling, A.J., Rangasamy, M., Worden, S.E., Narva, K.E., Siegfried, B.D., 2016. RNAi as a management tool for the western corn rootworm, *Diabrotica virgifera virgifera*. Pest Manag. Sci. 72, 1652–1663.
Gammon, D.B., Mello, C.C., 2015. RNA interference-mediated antiviral defense in insects. Curr. Opin. Insect Sci. 8, 111–120.
Ganbaatar, O., Cao, B., Zhang, Y., Bao, D., Bao, W., Wuriyanghan, H., 2017. Knockdown of *Mythimna separata* chitinase genes via bacterial expression and oral delivery of RNAi effectors. BMC Biotechnol. 17, 9.
Garbian, Y., Maori, E., Kalev, H., Shafir, S., Sela, I., 2012. Bidirectional transfer of RNAi between honey bee and *Varroa destructor*: *Varroa* gene silencing reduces *Varroa* population. PLoS Pathog. 8, e1003035.
Gillet, F.-X., Garcia, R.A., Macedo, L.L.P., Albuquerque, E.V.S., Silva, M.C.M., Grossi-de-Sa, M.F., 2017. Investigating engineered ribonucleoprotein particles to improve oral RNAi delivery in crop insect pests. Front. Physiol. 8, 256.
Goulson, D., Nicholls, E., Botias, C., Rotheray, E.L., 2015. Bee declines driven by combined stress from parasites, pesticides, and lack of flowers. Science 347, 10.
Guan, R., Li, H., Miao, X., 2017. RNAi pest control and enhanced BT insecticidal efficiency achieved by dsRNA of chymotrypsin-like genes in *Ostrinia furnacalis*. J. Pest Sci. 90, 745–757.
Guan, R.-B., Li, H.-C., Fan, Y.-J., Hu, S.-R., Christiaens, O., Smagghe, G., Miao, X.-X., 2018. A nuclease specific to lepidopteran insects suppresses RNAi. J. Biol. Chem. 293, 6011–6021.
Guedes, R.N.C., Smagghe, G., Stark, J.D., Desneux, N., 2016. Pesticide-induced stress in arthropod pests for optimized integrated pest management programs. Annu. Rev. Entomol. 61, 43–62.
Guo, Z., Kang, S., Chen, D., Wu, Q., Wang, S., Xie, W., Zhu, X., Baxter, S.W., Zhou, X., Jurat-Fuentes, J.L., Zhang, Y., 2015. MAPK signaling pathway alters expression of midgut ALP and ABCC genes and causes resistance to *Bacillus thuringiensis* Cry1Ac toxin in diamondback moth. PLoS Genet. 11, e1005124.
Heidel-Fischer, H.M., Vogel, H., 2015. Molecular mechanisms of insect adaptation to plant secondary compounds. Curr. Opin. Insect Sci. 8, 8–14.
Hu, J., Xia, Y., 2018. Increased virulence in the locust-specific fungal pathogen *Metarhizium acridum* expressing dsRNAs targeting the host F1F0–ATPase subunit genes. Pest Manag. Sci. https://doi.org/10.1002/ps.5085 (in press).
Jin, S., Singh, N.D., Li, L., Zhang, X., Daniell, H., 2015. Engineered chloroplast dsRNA silences cytochrome p450 monooxygenase, v-ATPase and chitin synthase genes in the insect gut and disrupts Helicoverpa armigera larval development and pupation. Plant Biotechnol. J. 13, 435–446.

Joga, M.R., Zotti, M.J., Smagghe, G., Christiaens, O., 2016. RNAi efficiency, systemic properties, and novel delivery methods for pest insect control: what we know so far. Front. Physiol. 7, 553.
Khajuria, C., Ivashuta, S., Wiggins, E., Flagel, L., Moar, W., Pleau, M., Miller, K., Zhang, Y., Ramaseshadri, P., Jiang, C., Hodge, T., Jensen, P., Chen, M., Gowda, A., McNulty, B., Vazquez, C., Bolognesi, R., Haas, J., Head, G., Clark, T., 2018. Development and characterization of the first dsRNA-resistant insect population from western corn rootworm, *Diabrotica virgifera virgifera* LeConte. PLoS One 13, e0197059.
Killiny, N., Hajeri, S., Tiwari, S., Gowda, S., Stelinski, L.L., 2014. Double-stranded RNA uptake through topical application, mediates silencing of five CYP4 genes and suppresses insecticide resistance in *Diaphorina citri*. PLoS One 9, e110536.
Kim, V.N., Han, J., Siomi, M.C., 2009. Biogenesis of small RNAs in animals. Nat. Rev. Mol. Cell Biol. 10, 126–139.
Kim, Y.H., Soumaila Issa, M., Cooper, A.M.W., Zhu, K.Y., 2015. RNA interference: applications and advances in insect toxicology and insect pest management. Pestic. Biochem. Physiol. 120, 109–117.
Kishk, A., Hijaz, F., Anber, H.A.I., AbdEl-Raof, T.K., El-Sherbeni, A.-H.D., Hamed, S., Killiny, N., 2017. RNA interference of acetylcholinesterase in the Asian citrus psyllid, *Diaphorina citri*, increases its susceptibility to carbamate and organophosphate insecticides. Pestic. Biochem. Physiol. 143, 81–89.
Kolliopoulou, A., Taning, C.N.T., Smagghe, G., Swevers, L., 2017. Viral delivery of dsRNA for control of insect agricultural pests and vectors of human disease: prospects and challenges. Front. Physiol. 8. https://doi.org/10.3389/fphys.2017.00399.
Lacey, L.A., Grzywacz, D., Shapiro-Ilan, D.I., Frutos, R., Brownbridge, M., Goettel, M.S., 2015. Insect pathogens as biological control agents: back to the future. J. Invertebr. Pathol. 132, 1–41.
Liu, J., Swevers, L., Iatrou, K., Huvenne, H., Smagghe, G., 2012. *Bombyx mori* DNA/RNA non-specific nuclease: expression of isoforms in insect culture cells, subcellular localization and functional assays. J. Insect Physiol. 58 (8), 1166–1176.
Liu, F., Huang, W., Wu, K., Qiu, Z., Huang, Y., Ling, E., 2017. Chapter seven—Exploiting innate immunity for biological pest control. Adv. Insect Physiol. 52, 199–230.
Luan, J.-B., Ghanim, M., Liu, S.-S., Czosnek, H., 2013. Silencing the ecdysone synthesis and signaling pathway genes disrupts nymphal development in the whitefly. Insect Biochem. Mol. Biol. 43, 740–746.
Maori, E., Garbian, Y., Kunik, V., Mozes-Koch, R., Malka, O., Kalev, H., Sabath, N., Sela, I., Shafir, S., 2018. A transmissible RNA pathway in honey bees. bioRxiv. https://doi.org/10.1101/299800.
Marques, J.T., Imler, J.-L., 2016. The diversity of insect antiviral immunity: insights from viruses. Curr. Opin. Microbiol. 32, 71–76.
Miranda, W., Paul, D., 2017. Gene silencing in non-model insects: overcoming hurdles using symbiotic bacteria for trauma-free sustainable delivery of RNA interference. Bioessays 39, 1600247.
Mitter, N., Worrall, E.A., Robinson, K.E., Li, P., Jain, R.G., Taochy, C., Fletcher, S.J., Carroll, B.J., Lu, G.Q., Xu, Z.P., 2017. Clay nanosheets for topical delivery of RNAi for sustained protection against plant viruses. Nat. Plants 3, 16207.
Ni, M., Ma, W., Wang, X., Gao, M., Dai, Y., Wei, X., Zhang, L., Peng, Y., Chen, S., Ding, L., Tian, Y., Li, J., Wang, H., Wang, X., Xu, G., Guo, W., Yang, Y., Wu, Y., Heuberger, S., Tabashnik, B., Zhang, T., & Zhu, Z., 2017. Next-generation transgenic cotton: pyramiding RNAi and Bt counters insect resistance. Plant Biotechnol. J. 15, 1204–1213.
Niu, J., Meeus, I., Cappelle, K., Piot, N., Smagghe, G., 2014. The immune response of the small interfering RNA pathway in the defense against bee viruses. Curr. Opin. Insect Sci. 6, 22–27.

Niu, J., Shen, G., Christiaens, O., Smagghe, G., He, L., Wang, J., 2018. Beyond insects: current status, achievements and future perspectives of RNAi in mite pests. Pest Manag. Sci. https://doi.org/10.1002/ps.5071 (in press).

Palli, S.R., 2014. RNA interference in Colorado potato beetle: steps toward development of dsRNA as a commercial insecticide. Curr. Opin. Insect Sci. 6, 1–8.

Ran, R., Li, T., Liu, X., Ni, H., Li, W., Meng, F., 2018. RNA interference-mediated silencing of genes involved in the immune responses of the soybean pod borer *Leguminivora glycinivorella* (Lepidoptera: Olethreutidae). PeerJ 6, e4931.

Remnant, E.J., Shi, M., Buchmann, G., Blacquière, T., Holmes, E.C., Beekman, M., Ashe, A., 2017. A diverse range of novel RNA viruses in geographically distinct honey bee populations. J. Virol. 91, e00158-00117.

Rundlöf, M., Andersson, G.K.S., Bommarco, R., Fries, I., Hederström, V., Herbertsson, L., Jonsson, O., Klatt, B.K., Pedersen, T.R., Yourstone, J., Smith, H.G., 2015. Seed coating with a neonicotinoid insecticide negatively affects wild bees. Nature 521, 77.

Schwenke, R.A., Lazzaro, B.P., Wolfner, M.F., 2016. Reproduction–immunity trade-offs in insects. Annu. Rev. Entomol. 61, 239–256.

Shi, M., Lin, X.-D., Tian, J.-H., Chen, L.-J., Chen, X., Li, C.-X., Qin, X.-C., Li, J., Cao, J.-P., Eden, J.-S., Buchmann, J., Wang, W., Xu, J., Holmes, E.C., Zhang, Y.-Z., 2016. Redefining the invertebrate RNA virosphere. Nature 540, 539–543.

Stahl, E., Hilfiker, O., Reymond, P., 2018. Plant–arthropod interactions: who is the winner? Plant J. 93, 703–728.

Stenberg, J.A., 2017. A conceptual framework for integrated pest management. Trends Plant Sci. 22, 759–769.

Sugahara, R., Tanaka, S., Jouraku, A., Shiotsuki, T., 2017. Geographic variation in RNAi sensitivity in the migratory locust. Gene 605, 5–11.

Swevers, L., Vanden Broeck, J., Smagghe, G., 2013. The possible impact of persistent virus infection on the function of the RNAi machinery in insects: a hypothesis. Front. Physiol. 4, 319.

Swevers, L., Liu, J., Smagghe, G., 2018. Defense mechanisms against viral infection in *Drosophila*: RNAi and non-RNAi. Viruses 10, 230.

Taning, C.N.T., Christiaens, O., Berkvens, N., Casteels, H., Maes, M., Smagghe, G., 2016. Oral RNAi to control *Drosophila suzukii*: laboratory testing against larval and adult stages. J. Pest Sci. 89, 803–814.

Taning, C.N.T., Christiaens, O., Li, X., Swevers, L., Casteels, H., Maes, M., Smagghe, G., 2018. Engineered Flock House Virus for targeted gene suppression through RNAi in fruit flies (*Drosophila melanogaster*) in vitro and in vivo. Front. Physiol. 9, 805.

Thakur, N., Upadhyay, S.K., Verma, P.C., Chandrashekar, K., Tuli, R., Singh, P.K., 2014. Enhanced whitefly resistance in transgenic tobacco plants expressing double stranded RNA of v-ATPase A gene. PLoS One 9, e87235.

Van Leeuwen, T., Dermauw, W., 2016. The molecular evolution of xenobiotic metabolism and resistance in chelicerate mites. Annu. Rev. Entomol. 61, 475–498.

Wang, M., Jin, H., 2017. Spray-induced gene silencing: a powerful innovative strategy for crop protection. Trends Microbiol. 25, 4–6.

Wang, K., Peng, Y., Pu, J., Fu, W., Wang, J., Han, Z., 2016. Variation in RNAi efficacy among insect species is attributable to dsRNA degradation in vivo. Insect Biochem. Mol. Biol. 77, 1–9.

Wang, M., Thomas, N., Jin, H., 2017a. Cross-kingdom RNA trafficking and environmental RNAi for powerful innovative pre- and post-harvest plant protection. Curr. Opin. Plant Biol. 38, 133–141.

Wang, R.L., Zhu-Salzman, K., Baerson, S.R., Xin, X.W., Li, J., Su, Y.J., Zeng, R.S., 2017b. Identification of a novel cytochrome P450 CYP321B1 gene from tobacco

cutworm (*Spodoptera litura*) and RNA interference to evaluate its role in commonly used insecticides. Insect Sci. 24, 235–247.
Webster, C.L., Waldron, F.M., Robertson, S., Crowson, D., Ferrari, G., Quintana, J.F., Brouqui, J.M., Bayne, E.H., Longdon, B., Buck, A.H., Lazzaro, B.P., Akorli, J., Haddrill, P.R., Obbard, D.J., 2015. The discovery, distribution, and evolution of viruses associated with *Drosophila melanogaster*. PLoS Biol. 13, 33.
Weiberg, A., Bellinger, M., Jin, H., 2015. Conversations between kingdoms: small RNAs. Curr. Opin. Biotechnol. 32, 207–215.
Xiao, Y., Liu, K., Zhang, D., Gong, L., He, F., Soberón, M., Bravo, A., Tabashnik, B.E., Wu, K., 2016. Resistance to *Bacillus thuringiensis* mediated by an ABC transporter mutation increases susceptibility to toxins from other bacteria in an invasive insect. PLoS Pathog. 12, e1005450.
Xu, L., Zhao, C.-Q., Xu, D.-J., Xu, G.-C., Xu, X.-L., Han, Z.-J., Zhang, Y.-N., Gu, Z.-Y., 2017. RNAi suppression of nuclear receptor genes results in increased susceptibility to sulfoxaflor in brown planthopper, *Nilaparvata lugens*. J. Asia Pac. Entomol. 20, 645–653.
Yoon, J.-S., Gurusamy, D., Palli, S.R., 2017. Accumulation of dsRNA in endosomes contributes to inefficient RNA interference in the fall armyworm, *Spodoptera frugiperda*. Insect Biochem. Mol. Biol. 90, 53–60.
Yu, X.D., Liu, Z.C., Huang, S.L., Chen, Z.Q., Sun, Y.W., Duan, P.F., Ma, Y.Z., Xia, L.Q., 2016. RNAi-mediated plant protection against aphids. Pest Manag. Sci. 72, 1090–1098.
Zhang, J., Khan, S.A., Hasse, C., Ruf, S., Heckel, D.G., Bock, R., 2015. Full crop protection from an insect pest by expression of long double-stranded RNAs in plastids. Science 347, 991–994.
Zhang, J., Khan, S.A., Heckel, D.G., Bock, R., 2017a. Next-generation insect-resistant plants: RNAi-mediated crop protection. Trends Biotechnol. 35, 871–882.
Zhang, C., Yan, S.-Q., Shen, B.-B., Ali, S., Wang, X.-M., Jin, F.-L., Cuthbertson, A.G.S., Qiu, B.-L., 2017b. RNAi knock-down of the Bemisia tabaci Toll gene (*BtToll*) increases mortality after challenge with destruxin A. Mol. Immunol. 88, 164–173.
Zotti, M., dos Santos, E.A., Cagliari, D., Christiaens, O., Taning, C.N.T., Smagghe, G., 2018. RNA interference technology in crop protection against arthropod pests, pathogens and nematodes. Pest Manag. Sci. 74, 1239–1250.

CHAPTER TWO

microRNAs as Regulators of Insect Host–Pathogen Interactions and Immunity

Sassan Asgari
Australian Infectious Diseases Research Centre, School of Biological Sciences, The University of Queensland, Brisbane, QLD, Australia

Contents

Abstract

Insects respond to infection through the stimulation of their immune pathways, which requires activation of transcription factors and changes in gene expression. Tight regulation of these pathways is essential for homeostasis and effective immune responses upon infection. As part of the RNA interference pathway, microRNAs play important roles in regulation of gene expression through their involvements in various biological processes, including immunity and host–pathogen interactions. Beside microRNAs encoded by the host, microRNAs or microRNA-like small noncoding RNAs produced by pathogens have also been implicated in regulating host as well as pathogen genes to provide a more suitable environment for pathogen replication and evasion of host defence responses. Here, after a brief introduction to microRNA biogenesis and the ways they interact with target genes, their role in the interaction of insects with a variety of pathogens is reviewed.

Advances in Insect Physiology, Volume 55
ISSN 0065-2806
https://doi.org/10.1016/bs.aiip.2018.07.004

1. INTRODUCTION

Noncoding RNAs (ncRNAs), whether small or long, have become the subject of intense research in the past decade with their functions realized in transcriptional and posttranscriptional regulation of gene expression. Among ncRNAs, microRNAs (miRNAs) with a typical size of ~22 nucleotides have attracted the most attention considering the variety of approaches that have been developed for their study. These include (1) next-generation sequencing platforms to profile small RNAs with their costs continuously declining; (2) bioinformatics tools to identify miRNAs and their precursor structures in genomic and transcriptomic data, their differential abundance, and potential target genes; and (3) experimental tools such as synthetic miRNA mimics (agomirs) and inhibitors (antagomirs) to oversupply or suppress the abundance of miRNAs, respectively, miRNA sponges to remove miRNAs, pull-down of miRNA biogenesis proteins (e.g. Argonaute pro teins) to validate miRNAs or determine the repertoire of their target genes, and in vitro validation of miRNA–target interaction using reporter genes.

miRNAs are produced in the majority of eukaryotes, some bacteria but also by a variety of viruses. The role of miRNAs in various biological processes has been documented, including development, immunity, cancer, response to environmental stimuli, and host–microorganism interactions (reviewed in Alvarez-Garcia and Miska, 2005; Bartel, 2009; Bushati and Cohen, 2007). As a consequence, dysregulation of miRNAs has been demonstrated to lead to abnormal cell functioning or development in certain instances to the extent that they are being considered as biomarkers in various clinical malignancies, in particular cancer (Mitchell et al., 2008; Pogribny, 2017; Reid et al., 2011). Similarly, miRNAs play various roles in different aspects of insect biology (reviewed in Asgari, 2013; Belles, 2017; Hussain and Asgari, 2015; Lucas and Raikhel, 2013). The focus of this review is on the role of miRNAs specifically in insect host–pathogen interactions.

2. HOW ARE miRNAs PRODUCED?

Typically, miRNA genes are transcribed similar to other mRNAs in the cell by RNA polymerase II in the form of primary miRNA (pri-miRNA) which is capped and polyadenylated (Fig. 1). The length of the pri-miRNA varies and contains one or more stem-loop structures that

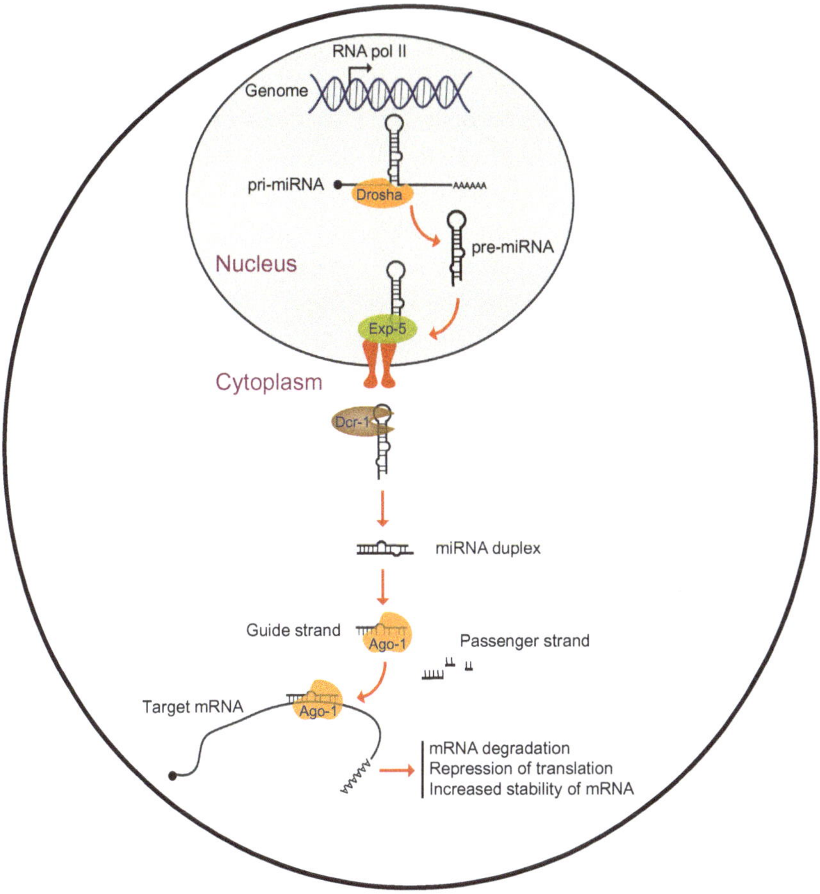

Fig. 1 Canonical microRNA biogenesis. miRNA genes are expressed by RNA polymerase II producing the primary miRNA (pri-miRNA) that contains one or more stem-loop structures. Similar to other cellular genes pri-miRNA has a polyA tail and a 5′ cap. The stem-loop is cleaved at the base by the ribonuclease enzyme Drosha producing the precursor miRNA (pre-miRNA) that is then transported into the cytoplasm by Exportin-5 (Exp-5). The hairpin head is then removed by another ribonuclease, Dicer-1 (Dcr-1), leading to formation of a miRNA duplex, which induces the formation of the RNA-induced silencing complex (RISC) in which Ago1 is the main component. The passenger strand is usually degraded, but the guide strand guides the RISC complex to target mRNA molecules binding to either the 5′UTR, open reading frame or the 3′UTR. The interaction may lead to mRNA degradation, repression of translation, and in certain instances increased stability of mRNA.

become substrates for the nuclear ribonuclease Drosha, which cleaves the stem-loop at the stem base in association with Pasha protein, releasing a hairpin structure of ~70 nt known as the precursor miRNA (pre-miRNA). pre-miRNAs may also be generated from introns of primary mRNA transcripts independent of Drosha through splicing, which are referred to as mirtrons (Flynt et al., 2010). pre-miRNA is subsequently transported into the cytoplasm with the assistance of Exportin-5 through the nuclear membrane. In the cytoplasm, the hairpin head is removed by another ribonuclease enzyme called Dicer-1. Unlike vertebrates, in insects, there are two Dicer proteins: Dicer-1 that is involved in miRNA biogenesis and Dicer-2 that participates in generation of short-interfering RNAs (Ye and Liu, 2008). The RNA duplex formed following the removal of the hairpin triggers the formation of the RNA-induced silencing complex (RISC), in which Argonaute 1 (Ago1) plays a major role. One of the strands of the loaded RNA duplex is degraded (the passenger strand) and the other strand, which is the mature miRNA, guides the RISC to target sequences based on sequence complementarity; hence, also known as the guide strand. Depending on whether the guide strand comes from the 5′ or the 3′ arm of the stem-loop, they are referred to as miR-X-5p or miR-X-3p, respectively, X being a designated miRNA number.

3. HOW DO miRNAs FUNCTION?

The main function of miRNAs is posttranscriptional regulation of gene expression by interacting with target mRNAs and hence fine-tuning protein levels (Bartel, 2009). However, beside their involvement in posttranscriptional regulation, growing evidence suggests that miRNAs present in the nucleus may interact with complementary sequences in the genome and cause induction or silencing of transcription depending on the context (reviewed in Zardo et al., 2012).

Complementarity of the miRNA nucleotides with those of the target sequences varies, but unlike plants, the complementarity is not usually 100%. Therefore, in animals, mismatches and bulges between miRNA and target sequences are expected. Nevertheless, complementarity of the seed region (nucleotides 2–8 from the 5′ end of the miRNA) is important for miRNA function; although extensive complementarity in the rest of the sequence may compensate for the lack of complete complementarity in the seed region (Lal et al., 2009; Wang, 2014). The strength of the interaction, in addition to the seed region complementarity, is assessed by the minimum free energy (mfe) that is an indicator of the strength of binding. Depending on the level of stringency, mfe should be lower than −20 kcal/mol.

While the majority of effective targets are within the 3′ untranslated region (UTR), miRNA binding may also occur in the 5′UTR or the open reading frame (ORF) (Bartel, 2009; Rigoutsos, 2009; Schnall-Levin et al., 2010). For example, using the cross-linking, ligation, and sequencing of hybrids technique, it was found that in human, 42.6% of targets reside in the coding region, 23.4% in the 3′UTR, and 3.8% in the 5′UTR (Helwak et al., 2013). In *Drosophila melanogaster*, the extent of biologically relevant targets in the coding regions was found to be as widespread as in the 3′UTRs. In addition, a considerable number of targets were found in the 5′UTRs (Schnall-Levin et al., 2010). Experimental evidence, however, suggests that the impact of 3′UTR target sites on gene regulation is higher than those in the 5′UTR or ORF in *D. melanogaster* (Agarwal et al., 2018). Besides, there might be multiple binding sites for the same or different miRNAs in different parts of a mRNA (Agarwal et al., 2018). In *Ae. aegypti*, an Ago1 cross-linking immunoprecipitation sequencing (CLIP-Seq) study showed a large number of potential targets in the coding sequences (Zhang et al., 2017), which further suggests that targets in ORFs could be common.

The consequence of miRNA–target interaction is usually degradation of mRNAs or repression of translation (Baek et al., 2008; Guo et al., 2010; Selbach et al., 2008). However, an increasing body of literature suggests that the interaction may in some cases lead to increases in the abundance of the target. In most instances the mechanism is not known but increase in the stability of mRNA, enhancement of transcription or translation have been reported as potential mechanisms (Bruno et al., 2011; He et al., 2016; Orom et al., 2008; Place et al., 2008). One reason for the presence of more examples of repression could be that investigators have basically searched for miRNA–target interactions with a bias towards repression.

4. miRNAs IN INSECTS

Similar to most other eukaryotes, miRNAs have been reported from a variety of different species of insects, in particular those with their complete genome sequence available. As the depth in sequencing platforms is increasing, ongoing research has led to revision or discovery of novel miRNAs. Insect miRNAs are involved in various aspects of insect biology, including development (Carthew et al., 2017; Chawla and Sokol, 2011), reproduction (Lucas and Raikhel, 2013), metamorphosis (Belles, 2017), memory formation (Busto et al., 2017), behaviour (Picao-Osorio et al., 2015), cast determination (Ashby et al., 2016), sexual dimorphism (Marco, 2014), metabolism and longevity (Gendron and Pletcher, 2017), insecticide resistance

(Guo et al., 2017), endosymbiosis (Feng et al., 2018), and host–pathogen interactions and immunity (Hussain and Asgari, 2014), among others. In this chapter, however, the focus is on miRNAs and their role in the interaction of insects as hosts with microorganisms, in particular the pathogenic ones.

5. REGULATORY ROLE OF miRNAs IN INSECT–PATHOGEN INTERACTIONS

Infection of an insect with a pathogen imposes immense physiological impacts on the organism at the cellular and organismal levels. Huge resources are allocated to inhibiting replication and spread of the pathogen through various immune defence mechanisms. This response requires rapid alterations in gene expression and modifications of proteins involved in the activation of immune pathways. As regulators of gene expression, miRNAs have also been implicated in insect responses to infection. Studies on various groups of pathogens, including viruses, bacteria, fungi, and protozoans, have shown alterations of the miRNA profiles of infected insects relative to the uninfected ones. While most of these alterations could be part of regulating immune responses, some could actually be induced by pathogens to facilitate their replication by compromising deference responses or providing resources required for their replication. Later, interaction of insects with different groups of pathogens is discussed.

5.1 Viruses

5.1.1 Baculoviruses

Baculoviruses are one of the most well-studied and utilized entomopathogenic viruses with applications in basic virology research, biological control of insect pests, expression vectors, gene therapy, and vaccine production (Airenne et al., 2013; Kost and Kemp, 2016; Moscardi, 1999). They are double-stranded DNA viruses belonging to the family *Baculoviridae*. During replication of the virus, two morphotypes of the virus are produced, the budded virus (BV), that is produced earlier in infection and is responsible for cell-to-cell infection; and the occlusion-derived virus (ODV) that is produced late in infection and is embedded in occlusion bodies responsible for horizontal transmission of the virus among host insects (Blissard and Rohrmann, 1990). The involvement of miRNAs in the interaction of baculoviruses with a number of insect hosts has been explored. In *Spodoptera frugiperda* (Sf9) cells, infection with Autographa californica multiple nucleopolyhedrovirus (AcMNPV) led to an overall reduction in cellular miRNAs when analysed 24 and 72 h postinfection (hpi) (Mehrabadi et al., 2013). However, a number of miRNAs, such as miR-184, miR-998, and miR-10, were more abundant at 24 hpi but their abundance declined at 72 hpi. Bantam that is one of the highly expressed miRNAs in insects and involved in various cellular and developmental processes was found to increase in abundance in Sf9 cells and *Spodoptera litura* larvae (Shi et al., 2016). Application of the inhibitor and the mimic of bantam in Sf9 cells led to changes in the expression of several AcMNPV genes; in particular, application of the bantam inhibitor led to 8, 10, and 40 times higher transcript levels of *left8*, *gp41*, and *p10* AcMNPV genes, respectively. Consequently, virus DNA replication increased significantly when bantam was inhibited (Shi et al., 2016). Conversely, application of bantam mimic led to reduction in virus DNA replication. The results imply that bantam may be involved in restricting AcMNPV's replication since its inhibition led to increase virus DNA replication.

Microarray analysis also showed differential expression of 15 *Helicoverpa armigera* miRNAs as a result of Helicoverpa armigera single nucleopolyhedrovirus (HaSNPV) infection (Jayachandran et al., 2013). miR-14, as one of the suppressed miRNAs upon HaSNPV infection, was found to positively regulate the ecdysone receptor in *H. armigera*. Oversupply of the miRNA in *H. armigera* through oral application of its mimic led to reduced virus DNA replication, suggesting that miR-14 suppression by the virus facilitates its replication. Reduction in miR-14 levels may contribute to the inhibition of moulting and pupation observed in baculovirus-infected larvae (O'reilly and Miller, 1989) by regulating the ecdysone receptor.

In *Bombyx mori* infected with Bombyx mori nucleopolyhedrovirus (BmNPV) 38 miRNAs were differentially expressed (Wu et al., 2016a). Target analysis of the differentially expressed miRNAs and their classification using Gene Ontology (GO) terms revealed their involvement in binding, catalytic, virion, and immune response to stimulus. One of the miRNAs that its abundance significantly increased after BmNPV infection was demonstrated to target the *DNA methyltransferase 2* (*Dnmt2*) gene. However, the role of the miRNA and its target gene in BmNPV–*B. mori* interaction remains to be investigated.

Beside the involvement of insect miRNAs in baculovirus–host interactions, several studies have reported identification and functional characterization of a number of baculovirus-encoded miRNAs that play a role in the interaction. Singh et al. identified eight miRNAs encoded by BmNPV (Singh et al., 2010). A follow-up study by the same group identified the host

GTP-binding nuclear protein Ran, which is a cofactor of Exportin-5, as the target gene of BmNPV-miR-1. Therefore, the host miRNA biogenesis is disturbed because transport of pre-miRNA to the cytoplasm is disrupted (Singh et al., 2012). As a consequence, in *B. mori* larvae the number of miRNAs declines in general after infection with BmNPV, which is similar to the observation in AcMNPV–Sf9 interaction (Mehrabadi et al., 2013). Inhibition of BmNPV-miR-1 led to higher expression levels of Ran and reduced virus replication, which suggests that interruption of the host miRNA machinery favours BmNPV replication, a strategy evolved in the virus to evade the host response by suppressing miRNAs that could have antiviral function. In fact, experimental evidence favours this assumption as the host bmo-miR-8, which targets the virus immediate early gene and therefore reduces virus replication, was shown to decline in abundance after viral infection (Singh et al., 2012). Another BmNPV-encoded miRNA, BmNPV-miR-3, was revealed to regulate the virus' own late genes, including P6.9 (Singh et al., 2014). This miRNA that is expressed in the early stages of virus infection negatively regulates the transcript levels of P6.9. When BmNPV-miR-3 was inhibited, virus replication significantly increased. Based on correlation of virus load with the expression levels of the antimicrobial protein Gloverin, the authors suggested that BmNPV-miR-3 reduces viral load by targeting P6.9 and other late genes at early stages of infection to avoid the host immune response towards the virus. Gloverin has been shown to be an antibaculoviral protein (Moreno-Habel et al., 2012).

miRNA acmnpv-miR-1 encoded by AcMNPV was shown to regulate replication of the virus by targeting the late genes *ODV-E25*, *ac18*, *ac94*, and *ac95* (Zhu et al., 2013, 2016). Considering that these genes are mostly required for the production of budded viruses, the authors suggested that reduction of these proteins later in infection results in decline in the production of BVs and viral DNA replication to favour formation of ODVs. In general, acmnpv-miR-1 restricts virus spread from cell to cell, but assists viral infection in larvae (Zhu et al., 2016). Ten miRNAs were validated to be encoded by Spodoptera litura nucleopolyderovirus when infecting Sf9 cells. The abundance of these miRNAs progressively increased by time. The GO classification of the predicted target genes of these virus-encoded miRNAs suggested their involvement in cell signalling, protein translation, and metabolic processes (Kharbanda et al., 2015).

In addition to the above-mentioned involvements of miRNAs in regulation of host or viral genes, based on other host–virus systems (see later), the possible role of miRNAs in covert infection of insects by baculoviruses has been suggested (Williams et al., 2017). However, this aspect needs to be experimentally investigated. Covert infection may contribute to the maintenance of the virus in field populations when the occurrence of horizontal transmission is low.

5.1.2 Cytoplasmic Polyhedrosis Virus

Cypoviruses are members of the family *Reoviridae* with a double-stranded, segmented RNA genome that produce virions encapsulated in polyhedra during infection, similar to baculoviruses (Mori and Metcalf, 2010). Cytoplasmic polyhedrosis viruses (CPVs) from the *Cypovirus* genus infect a variety of insect species. The most studied and economically important CPV is Bombyx mori cytoplasmic polyhedrosis virus (BmCPV) as it causes mortality in silkworm (Jiang and Xia, 2014). Infection of *B. mori* larvae by BmCPV was shown to lead to changes in the miRNA profile of the host midgut (Wu et al., 2013). The predicted target genes of the 58 differentially abundant miRNAs following infection were determined to be involved in stimulus and immune system process in silkworm. In a follow-up study, *insulin-related binding protein 2* (*IBP2*) gene was found as a target of miR-278-3p (Wu et al., 2016b). The abundance of miR-278-3p declines upon infection of *B. mori* larvae with BmCPV and leads to increases in the abundance of IBP2. Oversupply of the miRNA led to reduced *IBP2* levels and increased viral RNA transcription, which may imply that IBP2 could be involved in the host immune response to viral infection, since its reduction through miR-278-3p mimic treatment improved viral gene transcription. Further, miR-274-3p was found to directly target the *NS5* gene of BmCPV and reduce its abundance (Wu et al., 2017). Inhibition of the miRNA led to improved virus replication, suggesting that the miRNA might be induced upon infection as an antiviral mechanism to reduce virus replication.

BmCPV was also shown to encode putative miRNAs; BmCPV-miR-3 encoded by the third genomic RNA segment and BmCPV-miR-5 by the fifth genomic RNA segment (Pan et al., 2017). They were detected as early as 5 hpi. BmCPV-miR-3 was found to upregulate a virus gene (a structural protein gene on segment 3) but downregulate two host target genes (*purine nucleoside phosphorylase* and a *thrombospondin repeat protein 1-like protein*). However, the effect of this miRNA or the target genes on BmCPV replication was not investigated.

5.1.3 Mosquito-Borne Viruses

Mosquitoes are responsible for transmission of a number of viruses that cause mortality or morbidity in humans, including dengue virus (DENV), Zika virus (ZIKV), Chikungunya virus (CHIKV), and Yellow fever virus. These viruses are biologically transmitted, which means they replicate inside the mosquito prior to transmission. After taking a virus-contaminated blood meal, viruses infect the mosquito's midgut where they replicate and subsequently egress into the haemolymph where they circulate within the haemocoel and infect various other permissive tissues. Those that reach the salivary glands replicate in the glands and are transmitted to other hosts through subsequent blood-feeding events. This extrinsic period of incubation usually ranges from 7 to 14 days depending on the virus and the host. A number of studies have demonstrated the potential involvement of miRNAs during infection of mosquitoes by various arthropod-borne viruses (arboviruses). For more detailed reviews on this topic refer to these reviews (Asgari, 2014; Hussain et al., 2016).

In *Aedes aegypti and Aedes albopictus* mosquitoes, infections with ZIKV, DENV (serotype 2), and CHIKV have been shown to induce changes in the miRNA repertoire of the mosquito host (Campbell et al., 2014; Liu et al., 2016c; Maharaj et al., 2015; Saldaña et al., 2017; Shrinet et al., 2014; Su et al., 2017). This change appeared to be less pronounced in the case of DENV2 infection of the Aag2 cell line derived from *Ae. aegypti* embryos (Miesen et al., 2016), but infection of Aag2 cells with CHIKV led to changes in miRNAs, with the top three being miR-100, miR-2b, and miR-989 (Dubey et al., 2017). Overall, these studies suggest more common miRNAs altered in mosquitoes when infected with ZIKV and DENV2 (flaviviruses) as compared to those infected with CHIKV (an alphavirus) (Saldaña et al., 2017).

miR-281, a miRNA induced in the midgut of *Ae. albopictus* mosquitoes after DENV2 infection, was found to enhance DENV2 replication by interacting with the viral genomic 5′UTR (Zhou et al., 2014). Inhibition of the miRNA led to reduced DENV2 replication. A similar example exists in mammals when interaction of miR-122 in the liver with the 5′UTR of Hepatitis C virus (HCV) leads to enhanced viral replication by promoting translation of the genome (Jangra et al., 2010; Shimakami et al., 2012). Both DENV2 and HCV are positive sense RNA viruses. In contrast, miR-2b upregulated in *Ae. albopictus* mosquitoes infected with CHIKV was shown to be antiviral as inhibition of the miRNA led to enhanced CHIKV replication (Dubey et al., 2017). Based on the proposed model, miR-2b that is induced upon infection negatively targets *ubiquitin-related modified* gene leading to reduced tRNA thiolation in the cells that is implicated in viral replication. These examples show that host miRNAs may have proviral or antiviral functions by interacting with the virus genome or by regulating host target genes that would in turn affect viral replication.

In contrast to these pathogenic viruses, an insect-specific flavivirus (Palm Creek virus) did not cause major changes in the abundance of *Ae. aegypti* miRNAs (Lee et al., 2017). Lack of major changes in the case of the insect-specific flavivirus could be due to adaptation of the virus with mosquitoes as they are vertically transmitted and perhaps do not replicate to the same extent as pathogenic flaviviruses such as DENV. Virus infection may also alter posttranscriptional modification of miRNAs as was shown in *Ae. aegypti* mosquitoes infected with DENV (Etebari et al., 2015). These were modifications to the 5′ and 3′ ends and nucleotide substitutions; however, modifications to the 3′ end of miRNAs were more prevalent than those at the 5′ end.

5.1.4 Other Viruses

Ascoviruses are dsDNA viruses of the family *Ascoviridae* that cause fatal diseases in the larvae of several lepidopteran species (Bideshi et al., 2010). Research has shown that Heliothis virescens ascovirus 3e changes the miRNA profile of the host cells (Hussain and Asgari, 2010) and that a miRNA encoded by the ascovirus (HvAV-miR-1) fine-tunes replication of the virus by negatively regulating the transcripts of the viral DNA polymerase gene (Hussain et al., 2008). Helicoverpa zea nudivirus 1 also encodes two miRNAs from a noncoding RNA (the persistency-associated gene 1) that facilitates establishment of latent viral infection by targeting the viral early gene *hhi1* (Wu et al., 2011).

Amsacta moorei entomopoxvirus induces polyadenylation of host miRNAs by its encoded polyA polymerase. This makes the host miRNAs vulnerable to degradation by the host cell machinery and likely facilitates replication of the virus (Backes et al., 2012). In *D. melanogaster*, miR-8-5p was found to have a lower abundance in Drosophila C virus (DCV)-infected flies as compared with noninfected flies (Monsanto-Hearne et al., 2017a). The decrease in the abundance of the miRNA following infection leads to an increase in the transcript levels of its target gene, *cJun*, a transcription factor gene. cJun may in turn activate a number of genes that facilitate virus replication as its silencing led to reduced DCV replication. This provides another example of host miRNA manipulation by a virus to provide a more suitable environment for its replication. The abundance of another *D. melanogaster* miRNA, miR-956-3p, was also reduced following DCV infection (Monsanto-Hearne et al., 2017b). This reduction appears to delay DCV-induced mortality and accumulation of virions. The reduction in miR-956-3p leads to an increase in the abundance of its target gene, *Ectoderm-expressed 4* (*Ect4*) gene, which reduces virus replication and delays DCV-induced mortality. In turn, *Ect4* mutant flies died faster and accumulated more DCV virions. This suggests a miRNA-mediated regulation of a host gene that protects the host during DCV infection.

Involvement of miRNAs was implicated in infection of the adult bumble bee, *Bombus terrestis*, with slow bee paralysis virus (causing an avirulent infection) and Israeli acute paralysis virus (causing a virulent infection) as the bees miRNA profile was affected in comparison with uninfected bees (Niu et al., 2017). In addition, infection with slow bee paralysis virus led to increases in the expression of *Dicer-1* and *Ago1* genes. However, silencing of the *Dicer-1* gene did not have any effect on the genome copy number of the virus. The effect of plant viruses (Tomato yellow leaf curl China virus, a begomovirus, and rice black-streaked dwarf virus, a reovirus) on the miRNA profile of the host vector, the whitefly *Bemisia tabaci* and the small brown planthopper *Laodelphax striatellus*, respectively, has also been shown (Li et al., 2015; Wang et al., 2016).

5.2 Bacteria

Wolbachia pipientis is an endosymbiotic bacterium that is present in 40%–65% of insect species (Hilgenboecker et al., 2008; Zug and Hammerstein, 2012). It is mostly known for its effects on insect reproduction, in particular cytoplasmic incompatibility; uninfected females mating with infected males, or males and females with difference strains of *Wolbachia* mating produce no viable progeny (Bourtzis et al., 2003). Importantly, certain strains of *Wolbachia* inhibit replication of most RNA viruses, a phenomenon that is being utilized for reducing transmission of pathogenic viruses transmitted by mosquitoes (reviewed in Johnson, 2015a,b; Mcgraw and O'Neill, 2013). Various strains of *Wolbachia* stably transinfected into the mosquito *Ae. aegypti* have now been shown to block major arboviruses in the mosquito, such as DENV, ZIKV, and CHIKV (Aliota et al., 2016; Bian et al., 2010; Blagrove et al., 2012; Walker et al., 2011). Persistence of transinfected *Wolbachia* that does not naturally occur in *Ae. aegypti* requires adjustments both for the host and the endosymbiont. A microarray analysis of *w*MelPop-transinfected female *Ae. aegypti* showed alterations in the abundances of 13 mosquito miRNAs. Experimental work on manipulation of a number of these differentially abundant miRNAs demonstrated their importance in the maintenance of *Wolbachia*. For example, when the abundance of aae-miR-2940, which was found highly induced in *Wolbachia*-infected mosquitoes, was reduced by application of aae-miR-2940 inhibitor, *Wolbachia* density declined in mosquitoes (Hussain et al., 2011). Manipulation of three targets of the miRNA, *metalloprotease FtsH* (Hussain et al., 2011), *methyltransferase 2* (Zhang et al., 2013), and *arginine methyltransferase* (Zhang et al., 2014a), all led to reductions in *Wolbachia* density, suggesting the importance of this miRNA and regulation of its target genes for replication/persistence of *Wolbachia* in mosquito cells. In addition, inhibition of aae-miR-12, which is also induced in *w*MelPop-infected *Ae. aegypti* females, similarly led to reduction in *Wolbachia* density in cells persistently infected with the endosymbiont. Interestingly, no changes in the miRNA profile of a *D. melanogaster* cell line (JW18) infected with the *w*Mel strain of *Wolbachia* was observed, and a more recent study on Aag2 cell line infected with *w*Mel strain also showed no changes in the abundance of a selection of miRNAs tested by RT-qPCR (Thomas et al., 2018). There could be two reasons for the lack of change of miRNAs in these instances: (1) analysis of miRNAs in cell line vs whole organism and (2) differences between strains of *Wolbachia* (*w*Mel vs *w*MelPop, which is a more pathogenic strain and grows to higher densities). In support of the latter, aae-miR-2940 was also induced in *Ae. albopictus* C6/36 cells infected with *w*MelPop as compared to uninfected C6/36 cells (Hussain et al., 2011). Further, large changes in the abundance of aae-miR-2940 were observed in *w*MelPop and tetracycline-treated mosquitoes at the tissue level (Asad et al., 2018).

Although not an insect, in two-spotted spider mite, *Tetranychus urticae*, infected with *Wolbachia*, 91 miRNAs in females and 20 miRNAs in males were found differentially abundant as compared to the noninfected mites (Rong et al., 2014). The predicted targets were found to be mostly involved in sphingolipid metabolism, lysosome function, apoptosis, and lipid transporting in both males and females, as well as reproduction in females.

In addition to causing changes in the abundance of a number of *Ae. aegypti* miRNAs, *Wolbachia* *w*MelPop was also shown to affect the localization of miRNAs in the nucleus and the cytoplasm by changing the distribution of Ago1 between the two compartments (Hussain et al., 2013), but also modification of miRNA isomers that could lead to changes in the repertoire of their target genes (Mayoral et al., 2014). Further, miRNA-like small RNAs were detected from *Wolbachia w*MelPop that appear to be released from the endosymbiont into the cytoplasm with possible effects on host cell target genes. One of these *Wolbachia* miRNA-like small RNAs (WsnRNA-46) positively regulates *Dynein heavy chain* (*Dhc*) resulting in higher levels of *Dhc* in *Wolbachia*-infected mosquitoes as compared to noninfected ones. It has been shown that microtubules and its associated proteins Dynein and Dynactin are utilized by *Wolbachia* for localization and movements in *D. melanogaster* oocytes, and are important for normal *Wolbachia* levels and may play a role in its efficient maternal transmission (Ferree et al., 2005).

In addition to *Wolbachia*, as a 'facultative' endosymbiont, potential role of miRNAs in the interaction of an 'obligate' endosymbiont with its insect host has been investigated. *Buchnera aphidicola* is a gammaproteobacterium that has an obligate association with aphids playing essential roles in their hosts nutritional metabolism and development (Hansen and Moran, 2011; Moran and Yun, 2015; Wilson and Duncan, 2015). A recent study revealed that five miRNAs were highly expressed in the bacteriomes (the tissue where *Buchnera* replicates) of *Myzus persicae* aphids (Feng et al., 2018). These were miR-276, miR-3050, miR-10, miR-184a, and bantam. Comparison of miRNA profiles between bacteriomes and the gut, which does not harbour *Buchnera*, led to identification of 17 miRNAs that were differentially abundant between the tissues, including the five highly expressed miRNAs. Interestingly, apart from three miRNAs that were specific to *M. persicae*, the other 14 were also found to be differentially expressed across *M. persicae* lineages but also in another aphid species, *Aphis pisum*. The results suggest that these miRNAs are most likely involved in the interaction of the endosymbiont with their aphid hosts and their functions are potentially conserved. The investigators found that the orthologs of 10 of the 14 miRNAs have been implicated in insect host–pathogen interactions from various studies (Feng et al., 2018). Identification of the potential targets of these 14 miRNAs revealed overrepresentation of those genes involved in amino acid transport and metabolism, which is expected, as *Buchnera* provides essential and nonessential amino acids to the host (Moran and Yun, 2015). Furthermore, target gene profiling implicated a number of signal transduction pathways in *Buchnera*–aphid interaction that could perhaps facilitate symbiosis.

Beside endosymbiotic bacteria, the effect of *Bacillus thuringiensis* (Bt), as a pathogenic bacterium, has been explored in insects, but mainly in the context of resistance to Bt. In *Galleria mellonella*, exposure to Bt spores and endotoxins over 30 generations, which induced about 11-fold increase in resistance to Bt, led to differential expression of miRNAs resulting in the repression of potential genes that confer susceptibility to Bt (Mukherjee et al., 2017). In *Ostrinia furnacalis*, two independent studies found differential expression of miRNAs in Bt-resistant vs Bt-susceptible strains of the insect with several of them predicted to target Bt receptor genes, such as *cadherin* and *aminopeptidase N* (Xu et al., 2015; Yu et al., 2018). Feeding *Tribolium castaneum* beetles with the bacterial entomopathogen *Pseudomonas entomophila* and injection with peptidoglycan also altered the abundance of miRNAs in the beetles. Feeding *P. entomophila* to *T. castaneum* resulted in depletion of 11 and enrichment of 3 miRNAs (Freitak et al., 2012). In the case of injection of *T. castaneum* with peptidoglycan, 59 miRNAs were differentially abundant as compared to the control, with 21 miRNAs enriched. In *G. mellonella*, infection with *Listeria monocytogenes* altered the abundance of 90 miRNAs, with the predicted target genes involved in immunity, signal transduction, and autophagy (Mannala et al., 2017). Injection of honey bee, *Apis mellifera*, workers with *Serratia marcescens* for 6 h led to significant changes of miR-13a (enriched) and miR-1175 (depleted). miR-13a was predicted to target *dorsal-1* and *dorsal-2* of the Toll pathway (Lourenço et al., 2013), which suggests its potential involvement in manipulation of the host immune responses.

5.3 Fungi

In the western honey bee, *A. mellifera*, infection with *Nosema ceranae* spores induced changes in the miRNA profile of the midgut of newly emerged workers (Huang et al., 2015). Overall, 17 miRNAs showed differential expression across 6 days that correspond to the parasite's reproductive cycle. About 400 genes were predicted as potential targets of these differentially abundant miRNAs involved potentially in nine biological pathways based on Enzyme Code analysis of the target genes emphasizing the regulation of metabolism in *A. mellifera* (Huang et al., 2015). Interestingly, further analysis of the RNA-Seq data from the midgut of the infected *A. mellifera* workers revealed that this intracellular parasite also produces six miRNA-like small RNAs (Huang and Evans, 2016). However, the functional role of these small RNAs in regulation of host or the parasite genes has not been explored yet; although prediction tools identified host genes involved in metabolism as potential targets of *N. ceranae* small RNAs.

Different strains of *Metarhizium anisopliae* are widely used for the biocontrol of various insect pests. One of the key toxins produced by the fungus is Destruxin A that affects the host immune system (Gong et al., 2014; Pal et al., 2007). In *Plutella xylostella*, injection of the toxin into larvae led to differential expression of a number of miRNAs, including those involved in regulation of immunity, such as miR-263, miR-279, miR-306, miR-2a, and miR-308 (Shakeel et al., 2018). The target genes of these miRNAs were mostly determined to be involved in various immune pathways including Toll, IMD, and JAK–STAT pathways. miRNA-like small RNAs have also been identified from *M. anisopliae* using small RNA-Seq of different stages of the fungus (Zhou et al., 2012). However, it is not known if they play any role in the interaction of the fungus with its host insects. In *G. mellonella*, infection of larvae with *M. anisopliae* also resulted in changes in the abundance of miRNAs. In particular, miR-210b was significantly downregulated 4 and 9 days following infection. The transcript levels of one of its predicted target genes, a suppressor of immunity (NF-κB1), increased as a consequence suggesting miRNA-mediated suppression of the host immunity (Mukherjee and Vilcinskas, 2014).

5.4 Plasmodium

Plasmodium species are transmitted by *Anopheles* mosquitoes and cause malaria in humans and animals. This protozoan parasite goes through a number of different stages of development throughout its infection cycle in the mosquito vector. Following a blood meal from an infected individual and ingestion of the parasite in the form of gametocytes, the gametes form zygotes that transform into ookinets that are responsible for infecting the mosquito midgut cells (Beier, 1998). Suppression of the miRNA pathway by silencing *Dicer-1* and *Ago1* genes in *Anopheles gambiae* led to increased resistance to *Plasmodium falciparum*, suggesting that the pathway may facilitate infection by the parasite (Dennison et al., 2015). Further, microarray analysis of the midgut from *A. gambiae* mosquitoes 18 and 24h after feeding on a *P. falciparum*-infected blood meal showed increases in the abundance of aga-miR-989 and aga-miR-305 miRNAs. Inhibition of aae-miR-305 with artificial inhibitors led to increased resistance of *A. gambiae* to *P. falciparum* infection but also reduction in the gut microbiota. However, oversupply of the miRNA through miRNA mimics led to increased susceptibility of the mosquitoes to the parasite infection and expansion of the gut microbiota. The authors concluded that aga-miR-305 is an agonist of *P. falciparum* and midgut microbiota. Target analysis of the miRNA suggested that aga-miR-305 might target immune effector genes that are anti-*Plasmodium* (Dennison et al., 2015). The finding suggests a possible manipulation of a host miRNA by *P. falciparum* to suppress the host immune response to provide a more favourable environment for invasion of the midgut and subsequently other mosquito tissues. Interestingly, in an earlier study in *A. gambiae*, silencing of *Dicer-1* and *Ago1* genes led to increased susceptibility of the mosquitoes to *Plasmodium berghei* (Winter et al., 2007), suggesting that unlike *A. gambiae–P. falciparum* in *A. gambiae–P. berghei* interaction, miRNAs are involved in antimalarial defence. In another study, *A. gambiae* mosquitoes showed significant changes in the abundance of six miRNAs 3h after *P. berghei* infection (Biryukova et al., 2014). Four of these were novel miRNAs (miR-N3–N6), and two of them (miR-317, miR-2940) were conserved miRNAs which showed five- and threefold increases in their abundance upon infection, respectively. In this study, whole mosquitoes were used for sequencing.

In *A. anthropophagus*, feeding of mosquitoes on mice infected with *P. berghei* led to increases in the abundance of 13 and decreases in that of nine miRNAs in the midgut 48h after blood feeding (Liu et al., 2017). While miR-31 showed more than 10-fold increase, aga-miR-989 and aga-miR-305 were not among the miRNAs that showed elevated abundance. Among the nine depleted miRNAs in this study, six of them (miR-14, miR-2a, miR-92b, miR-957, miR-980, and miR-275) were also shown to have a reduction in abundance in *Anopheles stephansi* mosquitoes infected with *P. vinckei petteri* (Jain et al., 2014). In this study, a general reduction in the abundance of miRNAs following *Plasmodium* infection was observed. For a recent review on the role of miRNAs in *Anopheles* biology including interactions with *Plasmodium* refer to this review (Lampe and Levashina, 2017).

6. ROLE OF miRNAs IN IMMUNE REGULATION IN INSECTS

Homeostasis and prevention of immune hyperactivation is essential during noninfection periods. This requires tight regulation of genes involved in various immune pathways, such as the IMD and Toll pathways. As fine tuners of gene expression, miRNAs could play critical roles in the dynamics of immune responses and maintenance of immune homeostasis. For example, miR-8 was found to repress activation of antimicrobial peptide (AMP) genes for *Drosomycin* and *Diptericin* in *D. melanogaster* during noninfection period. In miR-8 knockout flies, the levels of the AMPs were significantly higher (Choi and Hyun, 2012). A follow-up study determined *Toll* (receptor), *Dorsal* (transcription factor), and *U-shaped* as targets of miR-8, suggesting negative regulation of multiple immune genes by one miRNA to maintain immune homeostasis (Lee and Hyun, 2014). A similar observation was made in *P. xylostella* larvae, in which miR-8 represses expression of the AMP gloverin during noninfection period (Etebari and Asgari, 2013). Experimental results in *P. xylostella* revealed that miR-8 positively regulates its target gene *serpin-27*, which is an inhibitor of the Toll pathway. However, upon parasitisation/infection, miR-8 levels decline resulting in reductions in serpin-27 and subsequent activation of the Toll pathway (Etebari and Asgari, 2013). Using a genome-wide approach in *D. melanogaster*, miR-310 family members (miR-310, miR-311, miR-312, and miR-313 expressed as a cluster) were also found to negatively regulate *Drosomycin* contributing to the homeostasis and dynamics of immune responses in the fly (Li et al., 2017a). The same research group also established that miR-958 targets *Toll* and *Dorsal-related immune factor* (*Dif*) as well leading to suppression of *Drosomycin* expression. Therefore, members of the miR-310 family (by targeting *Drosomycin* directly) and miR-958 (by targeting components of the Toll pathway) are all part of a regulatory network to control production of AMPs (Li et al., 2017b).

In *D. melanogaster*, availability of individual or cluster miRNA knockout mutants (Chen et al., 2014) has provided a great opportunity to examine their involvement in host–pathogen interactions and immunity. Using these mutants, it was shown that six miRNA knockout backgrounds affect the survival response and pathogen numbers in the flies infected with *Candida albicans* (Atilano et al., 2017); four miRNA (miR-317, miR-276b, miR-277-34, and miR-310-311-312-313) mutants were more susceptible to infection and two (miR-193 and miR-972-973-974) had greater survival as compared with the control. Interestingly, these mutants also exhibited dysregulation of *Drosomycin* and *immune-induced molecule 1* genes which are both immune effector molecules downstream of the Toll pathway. The study identified potential NF-κB binding sites in the proximity of 36% of *D. melanogaster* miRNA genes that could possibly be transcriptionally induced by immune signalling (Atilano et al., 2017).

miR-34 knockout *D. melanogaster* flies (Bejarano et al., 2012) were found defective in antibacterial defence, whereas overexpression of the miRNA led to a hyperactive immunity and improved clearance of *Escherichia coli* (Xiong et al., 2016). In the knockout flies, activation of the IMD signalling was compromised. This is in part achieved by repressing the septate junction protein *disks large Homologue 1* (*Dlg1*) and *Eip75B* (a transcription factor and component of ecdysone signalling) genes which both appear to be relevant to IMD signalling. Interestingly, miR-34 is regulated by ecdysone signalling through *cis*-regulatory genomic elements present in the miR-34 locus. These findings suggest an interplay of immunity and ecdysone signalling in the flies. Relevant to this topic, the steroid 20-hydroxy-ecdysone (20-HE) was shown to induce the conserved let-7 miRNA. Interestingly, 20-HE facilitates expression of the AMP *Dipterícin* mRNA (Meister and Richards, 1996), but let-7 represses the translation of *Diptericin*, suggesting that 20-HE plays a dual control role by inducing *Diptericin* transcription yet regulating its levels posttranscriptionally (Garbuzov and Tatar, 2010). Further, this instance provides another example of pleiotropic function of miRNAs in regulation of development and immunity as the role of let-7 in development has been well established (Caygill and Johnston, 2008; Sempere et al., 2002, 2003).

In *Manduca sexta*, immune induction by injection of *E. coli* and *Micrococcus luteus* and curdlan (a linear beta-1,3-glucan) led to changes in the miRNA profiles of hemocytes and fat body in the injected larvae as compared to control larvae (Zhang et al., 2014c). Predicted target genes of the differentially abundant miRNAs are involved in the classical immune pathways, the RNAi pathway, pattern recognition, and activation of prophenoloxidase important in melanisation.

7. CONCLUSIONS AND FUTURE PROSPECTS

In the context of insect host–pathogen interactions, infection usually results in alteration of the host miRNA profile, but its extent varies depending on the interaction. In earlier studies in which replicates of samples were not included in sequencing runs, the extent of variations reported were higher. However, with the costs of deep sequencing significantly declining, inclusion of replicates suggests that variations may not be as large. The altered miRNAs might be part of the host response to infection fine-tuning signalling pathways and immune responses, or conversely as a result of pathogen manipulation to provide an optimal environment for their replication. Pathogen-encoded miRNAs or miRNA-like small RNAs may also play a role in the interactions by suppressing the host antipathogen responses, facilitating replication or in some cases regulating replication of the pathogen to avoid over replication and rapid demise of the host. In that framework, an exciting area of research is the role of small noncoding RNAs in cross-kingdom and cell-to-cell communications in secreted forms such as exosomes. An unexplored related area is the role of miRNAs in the interaction of the gut microbiome with insects. This is an area that has recently been explored in mammals, suggesting that host miRNAs may shape the gut microbiota by entering bacteria and regulating bacterial gene transcripts and hence their growth (Liu et al., 2016a). It will not be surprising if similar associations exist in insects.

While studies of insect miRNAs are relatively recent, their applications in pest control or reduction of burden of diseases are possible. For example, inducible and regulated production of artificial miRNAs could be used to inhibit/reduce infection in beneficial insects, such as silkworm (Zhang et al., 2014b), to reduce economic losses due to infections. Vectors or the pathogens they carry could be targeted by overexpression of artificial pre-miRNAs to reduce their populations or specifically target the pathogens they might acquire through blood feeding (Liu et al., 2016b). Similarly, miRNAs could be utilized to enhance the efficacy of biocontrol agents used against agricultural pests or reduce pest populations by overproducing artificial miRNAs in genetically modified plants to target vital genes in pest insects leading to mortality or significant reductions in fecundity.

ACKNOWLEDGEMENTS

Work in Asgari's laboratory has been supported by the Australian Research Council (DP110102112, DP150101782) and National Health and Medical Research (APP1062983, APP1027110).

REFERENCES

Agarwal, V., Subtelny, A.O., Thiru, P., Ulitsky, I., Bartel, D.P., 2018. Predicting microRNA targeting efficacy in *Drosophila*. bioRxiv. https://doi.org/10.1101/198689.

Airenne, K.J., Hu, Y.C., Kost, T.A., Smith, R.H., Kotin, R.M., Ono, C., Matsuura, Y., Wang, S., Yla-Herttuala, S., 2013. Baculovirus: an insect-derived vector for diverse gene transfer applications. Mol. Ther. 21, 739–749.

Aliota, M.T., Peinado, S.A., Velez, I.D., Osorio, J.E., 2016. The *w*Mel strain of *Wolbachia* reduces transmission of Zika virus by *Aedes aegypti*. Sci. Rep. 6, 28792.

Alvarez-Garcia, I., Miska, E.A., 2005. MicroRNA functions in animal development and human disease. Development 132, 4653–4662.

Asad, S., Hussain, M., Hugo, L., Osei-Amo, S., Zhang, G., Watterson, D., Asgari, S., 2018. Suppression of the pelo protein by *Wolbachia* and its effect on dengue virus in *Aedes aegypti*. PLoS Negl. Trop. Dis. 12, e0006405.

Asgari, S., 2013. MicroRNA functions in insects. Insect Biochem. Mol. Biol. 43, 388–397.

Asgari, S., 2014. Role of microRNAs in arbovirus/vector interactions. Viruses 6, 3514–3534.

Ashby, R., Foret, S., Searle, I., Maleszka, R., 2016. MicroRNAs in honey bee caste determination. Sci. Rep. 6, 18794.

Atilano, M.L., Glittenberg, M., Monteiro, A., Copley, R.R., Ligoxygakis, P., 2017. MicroRNAs that contribute to coordinating the immune response in *Drosophila melanogaster*. Genetics 207, 163–178.
Backes, S., Shapiro, J., Sabin, L., Pham, A., Reyes, I., Moss, B., Cherry, S., Tenoever, B., 2012. Degradation of host microRNAs by poxvirus poly(A) polymerase reveals terminal RNA methylation as a protective antiviral mechanism. Cell Host Microbe 16, 200–210.
Baek, D., Villén, J., Shin, C., Camargo, F.D., Gygi, S.P., Bartel, D.P., 2008. The impact of microRNAs on protein output. Nature 455, 64–71.
Bartel, D.P., 2009. MicroRNAs: target recognition and regulatory functions. Cell 23, 215–233.
Beier, J.C., 1998. Malaria parasite development in mosquitoes. Annu. Rev. Entomol. 43, 519–543.
Bejarano, F., Bortolamiol-Becet, D., Dai, Q., Sun, K., Saj, A., Chou, Y.T., Raleigh, D.R., Kim, K., Ni, J.Q., Duan, H., Yang, J.S., Fulga, T.A., Van Vactor, D., Perrimon, N., Lai, E.C., 2012. A genome-wide transgenic resource for conditional expression of *Drosophila* microRNAs. Development 139, 2821–2831.
Belles, X., 2017. MicroRNAs and the evolution of insect metamorphosis. Annu. Rev. Entomol. 62, 111–125.
Bian, G., Xu, Y., Lu, P., Xie, Y., Xi, Z., 2010. The endosymbiotic bacterium *Wolbachia* induces resistance to dengue virus in *Aedes aegypti*. PLoS Pathog. 6, e1000833.
Bideshi, D.K., Bigot, Y., Federici, B.A., Spears, T., 2010. Asocviruses. In: Asgari, S., Johnson, K.N. (Eds.), Insect Virology. Caister Academic Press, Norfolk, UK.
Biryukova, I., Ye, T., Levashina, E., 2014. Transcriptome-wide analysis of microRNA expression in the malaria mosquito *Anopheles gambiae*. BMC Genomics 15, 557.
Blagrove, M.S.C., Arias-Goeta, C., Failloux, A.-B., Sinkins, S.P., 2012. *Wolbachia* strain wMel induces cytoplasmic incompatibility and blocks dengue transmission in *Aedes albopictus*. Proc. Natl. Acad. Sci. U. S. A. 102, 255–260.
Blissard, G.W., Rohrmann, G.F., 1990. Baculovirus diversity and molecular biology. Annu. Rev. Entomol. 35, 127–155.
Bourtzis, K., Braig, H.R., Karr, T.L., 2003. Cytoplasmic incompatibility. In: Bourtzis, K., Miller, T.A. (Eds.), Insect Symbiosis. CRC Press, FL, USA.
Bruno, I.G., Karam, R., Huang, L., Bhardwaj, A., Lou, C.H., Shum, E.Y., Song, H.W., Corbett, M.A., Gifford, W.D., Gecz, J., Pfaff, S.L., Wilkinson, M.F., 2011. Identification of a microRNA that activates gene expression by repressing nonsense-mediated RNA decay. Mol. Cell 42, 500–510.
Bushati, N., Cohen, S.M., 2007. microRNA functions. Annu. Rev. Cell Dev. Biol. 23, 175–205.
Busto, G.U., Guven-Ozkan, T., Davis, R.L., 2017. MicroRNA function in *Drosophila* memory formation. Curr. Opin. Neurobiol. 43, 15–24.
Campbell, C.L., Harrison, T., Hess, A.M., Ebel, G.D., 2014. MicroRNA levels are modulated in *Aedes aegypti* after exposure to dengue-2. Insect Mol. Biol. 23, 132–139.
Carthew, R.W., Agbu, P., Giri, R., 2017. MicroRNA function in *Drosophila melanogaster*. Semin. Cell Dev. Biol. 65, 29–37.
Caygill, E.E., Johnston, L.A., 2008. Temporal regulation of metamorphic processes in *Drosophila* by the let-7 and miR-125 heterochronic microRNAs. Curr. Biol. 18, 943–950.
Chawla, G., Sokol, N.S., 2011. MicroRNAs in *Drosophila* development. Int. Rev. Cell Mol. Biol. 286, 1–65.
Chen, Y.W., Song, S., Weng, R., Verma, P., Kugler, J.M., Buescher, M., Rouam, S., Cohen, S.M., 2014. Systematic study of *Drosophila* microRNA functions using a collection of targeted knockout mutations. Dev. Cell 31, 784–800.
Choi, I.K., Hyun, S., 2012. Conserved microRNA miR-8 in fat body regulates innate immune homeostasis in *Drosophila*. Dev. Comp. Immunol. 37, 50–54.

Dennison, N.J., Benmarzouk-Hidalgo, O.J., Dimopoulos, G., 2015. MicroRNA-regulation of *Anopheles gambiae* immunity to *Plasmodium falciparum* infection and midgut microbiota. Dev. Comp. Immunol. 49, 170–178.

Dubey, S.K., Shrinet, J., Jain, J., Ali, S., Sunil, S., 2017. *Aedes aegypti* microRNA miR-2b regulates ubiquitin-related modifier to control chikungunya virus replication. Sci. Rep. 7, 17666.

Etebari, K., Asgari, S., 2013. Conserved microRNA miR-8 blocks activation of the toll pathway by up-regulating serpin 27 transcripts. RNA Biol. 10, 1356–1364.

Etebari, K., Osei-Amo, S., Blomberg, S., Asgari, S., 2015. Dengue virus infection alters post-transcriptional modification of microRNAs in the mosquito vector *Aedes aegypti*. Sci. Rep. 5, 15968.

Feng, H., Wang, L., Wuchty, S., Wilson, A.C.C., 2018. microRNA regulation in an ancient obligate endosymbiosis. Mol. Ecol. 27, 1777–1793.

Ferree, P., Frydman, H., Li, J., Cao, J., Wieschaus, E., Sullivan, W., 2005. *Wolbachia* utilizes host microtubules and dynein for anterior localization in the *Drosophila* oocyte. PLoS Pathog. 1, e14.

Flynt, A.S., Greimann, J.C., Chung, W.J., Lima, C.D., Lai, E.C., 2010. MicroRNA biogenesis via splicing and exosome-mediated trimming in *Drosophila*. Mol. Cell 38, 900–907.

Freitak, D., Knorr, E., Vogel, H., Vilcinskas, A., 2012. Gender- and stressor-specific microRNA expression in *Tribolium castaneum*. Biol. Lett. 8, 860–863.

Garbuzov, A., Tatar, M., 2010. Hormonal regulation of *Drosophila* microRNA let-7 and miR-125 that target innate immunity. Fly 4, 4.

Gendron, C.M., Pletcher, S.D., 2017. MicroRNAs mir-184 and let-7 alter *Drosophila* metabolism and longevity. Aging Cell 16, 1434–1438.

Gong, L., Chen, X., Liu, C., Jin, F., Hu, Q., 2014. Gene expression profile of *Bombyx mori* hemocyte under the stress of destruxin A. PLoS One 9, e96170.

Guo, H., Ingolia, N.T., Weissman, J.S., Bartel, D.P., 2010. Mammalian microRNAs predominantly act to decrease target mRNA levels. Nature 466, 835–840.

Guo, Q., Huang, Y., Zou, F., Liu, B., Tian, M., Ye, W., Guo, J., Sun, X., Zhou, D., Sun, Y., Ma, L., Shen, B., Zhu, C., 2017. The role of miR-2∼13∼71 cluster in resistance to deltamethrin in *Culex pipiens* pallens. Insect Biochem. Mol. Biol. 84, 15–22.

Hansen, A.K., Moran, N.A., 2011. Aphid genome expression reveals host-symbiont cooperation in the production of amino acids. Proc. Natl. Acad. Sci. U. S. A. 108, 2849–2854.

He, J., Chen, Q., Wei, Y., Jiang, F., Yang, M., Hao, S., Guo, X., Chen, D., Kang, L., 2016. MicroRNA-276 promotes egg-hatching synchrony by up-regulating brm in locusts. Proc. Natl. Acad. Sci. U. S. A. 113, 584–589.

Helwak, A., Kudla, G., Dudnakova, T., Tollervey, D., 2013. Mapping the human miRNA interactome by CLASH reveals frequent noncanonical binding. Cell 153, 654–665.

Hilgenboecker, K., Hammerstein, P., Schlattmann, P., Telschow, A., Werren, J.H., 2008. How many species are infected with *Wolbachia*?—a statistical analysis of current data. FEMS Microbiol. Lett. 218, 215–220.

Huang, Q., Evans, J.D., 2016. Identification of microRNA-like small RNAs from fungal parasite *Nosema ceranae*. J. Invertebr. Pathol. 133, 107–109.

Huang, Q., Chen, Y., Wang, R.W., Schwarz, R.S., Evans, J.D., 2015. Honey bee microRNAs respond to infection by the microsporidian parasite *Nosema ceranae*. Sci. Rep. 5, 17494.

Hussain, M., Asgari, S., 2010. Functional analysis of a cellular microRNA in insect host-ascovirus interaction. J. Virol. 84, 612–620.

Hussain, M., Asgari, S., 2014. MicroRNAs as mediators of insect host–pathogen interactions and immunity. J. Insect Physiol. 70, 151–158.

Hussain, M., Asgari, S., 2015. A glance on the role of miRNAs in insect life. In: Hoffmann, K.H. (Ed.), Insect Molecular Biology and Ecology. CRC Press.

Hussain, M., Taft, R.J., Asgari, S., 2008. An insect virus-encoded microRNA regulates viral replication. J. Virol. 82, 9164–9170.
Hussain, M., Frentiu, F.D., Moreira, L.A., O'neill, S.L., Asgari, S., 2011. *Wolbachia* utilizes host microRNAs to manipulate host gene expression and facilitate colonization of the dengue vector *Aedes aegypti*. Proc. Natl. Acad. Sci. U. S. A. 108, 9250–9255.
Hussain, M., O'neill, S.L., Asgari, S., 2013. *Wolbachia* interferes with the intracellular distribution of Argonaute 1 in the dengue vector *Aedes aegypti* by manipulating the host microRNAs. RNA Biol. 10, 1868–1875.
Hussain, M., Etebari, K., Asgari, S., 2016. Functions of small RNAs in mosquitoes. Adv. Insect Physiol. 51, 189–222.
Jain, S., Rana, V., Shrinet, J., Sharma, A., Tridibes, A., Sunil, S., Bhatnagar, R., 2014. Blood feeding and *Plasmodium* infection alters the miRNome of *Anopheles stephensi*. PLoS One 9, e98402.
Jangra, R.K., Yi, M., Lemon, S.M., 2010. Regulation of hepatitis C virus translation and infectious virus production by the microRNA miR-122. J. Virol. 84, 6615–6625.
Jayachandran, B., Hussain, M., Asgari, S., 2013. Regulation of *Helicoverpa armigera* ecdysone receptor by miR-14 and its potential link to baculovirus infection. J. Invertebr. Pathol. 114, 151–157.
Jiang, L., Xia, Q., 2014. The progress and future of enhancing antiviral capacity by transgenic technology in the silkworm *Bombyx mori*. Insect Biochem. Mol. Biol. 48, 1–7.
Johnson, K.N., 2015a. Bacteria and antiviral immunity in insects. Curr. Opin. Insect Sci. 8, 97–103.
Johnson, K.N., 2015b. The impact of *Wolbachia* on virus infection in mosquitoes. Viruses 7, 5705–5717.
Kharbanda, N., Jalali, S.K., Ojha, R., Bhatnagar, R.K., 2015. Temporal expression profiling of novel Spodoptera litura nucleopolyhedrovirus-encoded microRNAs upon infection of Sf21 cells. J. Gen. Virol. 96, 688–700.
Kost, T.A., Kemp, C.W., 2016. Fundamentals of baculovirus expression and applications. Adv. Exp. Med. Biol. 896, 187–197.
Lal, A., Navarro, F., Maher, C.A., Maliszewski, L.E., Yan, N., O'day, E., Chowdhury, D., Dykxhoorn, D.M., Tsai, P., Hofmann, O., Becker, K.G., Gorospe, M., Hide, W., Liebeman, J., 2009. miR-24 inhibits cell proliferation by targeting E2F2, MYC, and other cell-cycle genes via binding to "seedless" 3'UTR microRNA recognition elements. Mol. Cell 35, 610–625.
Lampe, L., Levashina, E.A., 2017. The role of microRNAs in *Anopheles* biology-an emerging research field. Parasite Immunol. 39, e12405.
Lee, G., Hyun, S., 2014. Multiple targets of the microRNA miR-8 contribute to immune homeostasis in *Drosophila*. Dev. Comp. Immunol. 45, 245–251.
Lee, M., Etebari, K., Hall-Mendelin, S., Van Den Hurk, A.F., Hobson-Peters, J., Vatipally, S., Schnettler, E., Hall, R., Asgari, S., 2017. Understanding the role of microRNAs in the interaction of *Aedes aegypti* mosquitoes with an insect-specific flavivirus. J. Gen. Virol. 98, 892–1903.
Li, J.M., Zhou, Y.R., Sun, Z.T., Wang, X., Xie, L., Chen, J.P., 2015. Identification and profiling of conserved and novel microRNAs in *Laodelphax striatellus* in response to rice black-streaked dwarf virus (RBSDV) infection. Genom. Data 3, 63–69.
Li, S., Li, Y., Shen, L., Jin, P., Chen, L., Ma, F., 2017a. miR-958 inhibits toll signaling and Drosomycin expression via direct targeting of toll and Dif in *Drosophila melanogaster*. Am. J. Physiol. Cell Physiol. 312, C103–C110.
Li, Y., Li, S., Li, R., Xu, J., Jin, P., Chen, L., Ma, F., 2017b. Genome-wide miRNA screening reveals miR-310 family members negatively regulate the immune response in *Drosophila melanogaster* via co-targeting Drosomycin. Dev. Comp. Immunol. 68, 34–45.

Liu, S., Da Cunha, A.P., Rezende, R.M., Cialic, R., Wei, Z., Bry, L., Comstock, L.E., Gandhi, R., Weiner, H.L., 2016a. The host shapes the gut microbiota via fecal micro-RNA. Cell Host Microbe 19, 32–43.

Liu, Y.X., Li, F.X., Liu, Z.Z., Jia, Z.R., Zhou, Y.H., Zhang, H., Yan, H., Zhou, X.Q., Chen, X.G., 2016b. Integrated analysis of miRNAs and transcriptomes in *Aedes albopictus* midgut reveals the differential expression profiles of immune-related genes during dengue virus serotype-2 infection. Insect Sci. 23, 377–385.

Liu, P., Li, X., Gu, J., Dong, Y., Liu, Y., Santhosh, P., Chen, X., 2016c. Development of non-defective recombinant densovirus vectors for microRNA delivery in the invasive vector mosquito, *Aedes albopictus*. Sci. Rep. 6, 20979.

Liu, W., Hao, Z., Huang, L., Chen, L., Wei, Q., Cai, L., Liang, S., 2017. Comparative expression profile of microRNAs in *Anopheles anthropophagus* midgut after blood-feeding and *Plasmodium* infection. Parasit. Vectors 10, 86.

Lourenço, A.P., Guidugli-Lazzarini, K.R., Freitas, F.C., Bitondi, M.M., Simões, Z.L., 2013. Bacterial infection activates the immune system response and dysregulates microRNA expression in honey bees. Insect Biochem. Mol. Biol. 43, 474–482.

Lucas, K., Raikhel, A.S., 2013. Insect microRNAs: biogenesis, expression profiling and biological functions. Insect Biochem. Mol. Biol. 43, 24–38.

Maharaj, P., Widen, S., Huang, J., Wood, T., Thangamani, S., 2015. Discovery of mosquito saliva microRNAs during CHIKV infection. PLoS Negl. Trop. Dis. 9, e0003386.

Mannala, G.K., Izar, B., Rupp, O., Schultze, T., Goesmann, A., Chakraborty, T., Hain, T., 2017. Listeria monocytogenes induces a virulence-dependent microRNA signature that regulates the immune response in *Galleria mellonella*. Front. Microbiol. 8, 2463.

Marco, A., 2014. Sex-biased expression of microRNAs in *Drosophila melanogaster*. Open Biol. 4, 140024.

Mayoral, J., Etebari, K., Hussain, M., Khromykh, A., Asgari, S., 2014. *Wolbachia* infection modifies the profile, shuttling and structure of microRNAs in a mosquito cell line. PLoS One 9, e96107.

Mcgraw, E.A., O'Neill, S.L., 2013. Beyond insecticides: new thinking on an ancient problem. Nat. Rev. Microbiol. 11, 181–193.

Mehrabadi, M., Hussain, M., Asgari, S., 2013. MicroRNAome of *Spodoptera frugiperda* cells (Sf9) and its alteration following baculovirus infection. J. Gen. Virol. 94, 1385–1397.

Meister, M., Richards, G., 1996. Ecdysone and insect immunity: the maturation of the inducibility of the diptericin gene in *Drosophila* larvae. Insect Biochem. Mol. Biol. 26, 155–160.

Miesen, P., Ivens, A., Buck, A., Van Rij, R., 2016. Small RNA profiling in dengue virus 2-infected *Aedes* mosquito cells reveals viral piRNAs and novel host miRNAs. PLoS Negl. Trop. Dis. 10, e0004452.

Mitchell, P.S., Parkin, R.K., Kroh, E.M., Fritz, B.R., Wyman, S.K., Pogosova-Agadjanyan, E.L., Peterson, A., Noteboom, J., O'briant, K.C., Allen, A., Lin, D.W., Urban, N., Drescher, C.W., Knudsen, B.S., Stirewalt, D.L., Gentleman, R., Vessella, R.L., Nelson, P.S., Martin, D.B., Tewari, M., 2008. Circulating microRNAs as stable blood-based markers for cancer detection. Proc. Natl. Acad. Sci. U. S. A. 105, 10513–10518.

Monsanto-Hearne, V., Asad, S., Asgari, S., Johnson, K.N., 2017a. *Drosophila* microRNA modulates viral replication by targeting a homologue of mammalian cJun. J. Gen. Virol. 98, 1904–1912.

Monsanto-Hearne, V., Tham, A.L., Wong, Z.S., Asgari, S., Johnson, K.N., 2017b. Drosophila miR-956 suppression modulates ectoderm-expressed 4 and inhibits viral replication. Virology 502, 20–27.

Moran, N.A., Yun, Y., 2015. Experimental replacement of an obligate insect symbiont. Proc. Natl. Acad. Sci. U. S. A. 112, 2093–2096.

Moreno-Habel, D., Biglang-Awa, I., Dulce, A., Luu, D., Garcia, P., Weers, P., Haas-Stapleton, E., 2012. Inactivation of the budded virus of *Autographa californica* M nucleopolyhedrovirus by gloverin. J. Invertebr. Pathol. 110, 92–101.

Mori, H., Metcalf, P., 2010. Cypoviruses. In: Asgari, S., Johnson, K.N. (Eds.), Insect Virology. Caister Academic Press, Great Britain.

Moscardi, F., 1999. Assessment of the application of baculoviruses for control of Lepidoptera. Annu. Rev. Entomol. 44, 257–289.

Mukherjee, K., Vilcinskas, A., 2014. Development and immunity-related microRNAs of the lepidopteran model host *Galleria mellonella*. BMC Genomics 15, 705.

Mukherjee, K., Grizanova, E., Chertkova, E., Lehmann, R., Dubovskiy, I., Vilcinskas, A., 2017. Experimental evolution of resistance against *Bacillus thuringiensis* in the insect model host *Galleria mellonella* results in epigenetic modifications. Virulence 8, 1618–1630.

Niu, J., Meeus, I., De Coninck, D.I., Deforce, D., Etebari, K., Asgari, S., Smagghe, G., 2017. Infections of virulent and avirulent viruses differentially influenced the expression of dicer-1, ago-1, and microRNAs in *Bombus terrestris*. Sci. Rep. 7, 45620.

O'reilly, D.R., Miller, L.K., 1989. A baculovirus blocks insect molting by producing ecdysteroid UDP-glucosyl transferase. Science 245, 1110–1112.

Orom, U.A., Nielsen, F.C., Lund, A.H., 2008. MicroRNA-10a binds the 5′UTR of ribosomal protein mRNAs and enhances their translation. Mol. Cell 30, 460–471.

Pal, S., St Leger, R.J., Wu, L.P., 2007. Fungal peptide destruxin A plays a specific role in suppressing the innate immune response in *Drosophila melanogaster*. J. Biol. Chem. 282, 8969–8977.

Pan, Z.H., Wu, P., Gao, K., Hou, C.X., Qin, G.X., Geng, T., Guo, X.J., 2017. Identification and characterization of two putative microRNAs encoded by Bombyx mori cypovirus. Virus Res. 233, 86–94.

Picao-Osorio, J., Johnston, J., Landgraf, M., Berni, J., Alonso, C.R., 2015. MicroRNA-encoded behavior in Drosophila. Science 350, 815–820.

Place, R.F., Li, L.C., Pookot, D., Noonan, E.J., Dahiya, R., 2008. MicroRNA-373 induces expression of genes with complementary promoter sequences. Proc. Natl. Acad. Sci. U. S. A. 105, 1608–1613.

Pogribny, I.P., 2017. MicroRNAs as biomarkers for clinical studies. Exp. Biol. Med. 243, 283–290.

Reid, G., Kirschner, M.B., Van Zandwijk, N., 2011. Circulating microRNAs: association with disease and potential use as biomarkers. Crit. Rev. Oncol. Hematol. 80, 193–208.

Rigoutsos, I., 2009. New tricks for animal microRNAs: targeting of amino acid coding regions at conserved and nonconserved sites. Cancer Res. 69, 3245–3248.

Rong, X., Zhang, Y.K., Zhang, K.J., Hong, X.Y., 2014. Identification of *Wolbachia*-responsive microRNAs in the two-spotted spider mite, *Tetranychus urticae*. BMC Genomics 15, 1122.

Saldaña, M.A., Etebari, K., Hart, C.E., Widen, S.G., Wood, T.G., Thangamani, S., Asgari, S., Hughes, G.L., 2017. Zika virus alters the microRNA expression profile and elicits an RNAi response in *Aedes aegypti* mosquitoes. PLoS Negl. Trop. Dis. 11, e0005760.

Schnall-Levin, M., Zhao, Y., Perrimon, N., Berger, B., 2010. Conserved microRNA targeting in *Drosophila* is as widespread in coding regions as in 3'UTRs. Proc. Natl. Acad. Sci. U. S. A. 107, 15751–15756.

Selbach, M., Schwanhäusser, B., Thierfelder, N., Fang, Z., Khanin, R., Rajewsky, N., 2008. Widespread changes in protein synthesis induced by microRNAs. Nature 455, 58–63.

Sempere, L.F., Dubrovsky, E.B., Dubrovskaya, V.A., Berger, E.M., Ambros, V., 2002. The expression of the let-7 small regulatory RNA is controlled by ecdysone during metamorphosis in *Drosophila melanogaster*. Dev. Biol. 244, 170–179.

Sempere, L.F., Sokol, N.S., Dubrovsky, E.B., Berger, E.M., Ambros, V., 2003. Temporal regulation of microRNA expression in *Drosophila melanogaster* mediated by hormonal signals and broad-complex gene activity. Dev. Biol. 259, 9–18.

Shakeel, M., Xu, X., Xu, J., Li, S., Yu, J., Zhou, X., Xu, X., Hu, Q., Yu, X., Jin, F., 2018. Genome-wide identification of destruxin A-responsive immunity-related microRNAs in diamondback moth, *Plutella xylostella*. Front. Immunol. 9, 185.

Shi, X., Ran, Z., Li, S., Yin, J., Zhong, J., 2016. The effect of microRNA bantam on baculovirus AcMNPV infection in vitro and in vivo. Viruses 8, E136.

Shimakami, T., Yamane, D., Jangra, R.K., Kempf, B.J., Spaniel, C., Barton, D.J., Lemon, S.M., 2012. Stabilization of hepatitis C virus RNA by an Ago2-miR-122 complex. Proc. Natl. Acad. Sci. U. S. A. 109, 941–946.

Shrinet, J., Jain, S., Jain, J., Bhatnagar, R., Sunil, S., 2014. Next generation sequencing reveals regulation of distinct *Aedes* microRNAs during chikungunya virus development. PLoS Negl. Trop. Dis. 8, e2616.

Singh, J., Singh, C.P., Bhavani, A., Nagaraju, J., 2010. Discovering microRNAs from *Bombyx mori* nucleopolyhedrosis virus. Virology 407, 120–128.

Singh, C.P., Singh, J., Nagaraju, J., 2012. A baculovirus-encoded microRNA (miRNA) suppresses its host miRNA biogenesis by regulating the exportin-5 cofactor Ran. J. Virol. 86, 7867–7879.

Singh, C.P., Singh, J., Nagaraju, J., 2014. Bmnpv-miR-3 facilitates BmNPV infection 1 by modulating the 2 expression of viral P6.9 and other late genes in *Bombyx mori*. Insect Biochem. Mol. Biol. 49, 59–69.

Su, J., Li, C., Zhang, Y., Yan, T., Zhu, X., Zhao, M., Xing, D., Dong, Y., Guo, X., Zhao, T., 2017. Identification of microRNAs expressed in the midgut of *Aedes albopictus* during dengue infection. Parasit. Vectors 10, 63.

Thomas, S., Verma, J., Woolfit, M., O'neill, S.L., 2018. *Wolbachia*-mediated virus blocking in mosquito cells is dependent on XRN1-mediated viral RNA degradation and influenced by viral replication rate. PLoS Pathog. 14, e1006879.

Walker, T., Johnson, P.H., Moreira, L.A., Iturbe-Ormaetxe, I., Frentiu, F.D., Mcmeniman, C.J., Leong, Y.S., Dong, Y., Axford, J., Kriesner, P., Lloyd, A.L., Ritchie, S.A., O'neill, S.L., Hoffman, A.A., 2011. The *w*Mel *Wolbachia* strain blocks dengue and invades caged *Aedes aegypti* populations. Nature 476, 450–453.

Wang, X., 2014. Composition of seed sequence is a major determinant of microRNA targeting patterns. Bioinformatics 30, 1377–1383.

Wang, B., Wang, L., Chen, F., Yang, X., Ding, M., Zhang, Z., Liu, S.S., Wang, X.W., Zhou, X., 2016. MicroRNA profiling of the whitefly *Bemisia tabaci* Middle East-Asia minor I following the acquisition of tomato yellow leaf curl China virus. Virol. J. 13, 20.

Williams, T., Virto, C., Murillo, R., Caballero, P., 2017. Covert infection of insects by baculoviruses. Front. Microbiol. 8, 1337.

Wilson, A.C., Duncan, R.P., 2015. Signatures of host/symbiont genome coevolution in insect nutritional endosymbioses. Proc. Natl. Acad. Sci. U. S. A. 112, 10255–10261.

Winter, F., Edaye, S., Hüttenhofer, A., Brunel, C., 2007. *Anopheles gambiae* miRNAs as actors of defence reaction against *Plasmodium* invasion. Nucleic Acids Res. 35, 6953–6962.

Wu, Y.-L., Wu, C.P., Liu, C.Y.Y., Hsu, P.W.-C., Wu, E.C., Chao, Y.C., 2011. A noncoding RNA of insect HzNV-1 virus establishes latent viral infection through microRNA. Sci. Rep. 1, 60.

Wu, P., Han, S., Chen, T., Qin, G., Li, L., Guo, X., 2013. Involvement of microRNAs in infection of silkworm with *Bombyx mori* cytoplasmic polyhedrosis virus (BmCPV). PLoS One 8, e68209.

Wu, P., Jiang, X., Guo, X., Li, L., Chen, T., 2016a. Genome-wide analysis of differentially expressed microRNA in *Bombyx mori* infected with nucleopolyhedrosis virus. PLoS One 11, e0165865.

Wu, P., Qin, G., Qian, H., Chen, T., Guo, X., 2016b. Roles of miR-278-3p in IBP2 regulation and Bombyx mori cytoplasmic polyhedrosis virus replication. Gene 575, 264–269.

Wu, P., Jiang, X., Sang, Q., Annan, E., Cheng, T., Guo, X., 2017. Inhibition of miR-274-3p increases BmCPV replication by regulating the expression of BmCPV NS5 gene in *Bombyx mori*. Virus Genes 53, 643–649.

Xiong, X.P., Kurthkoti, K., Chang, K.Y., Li, J.L., Ren, X., Ni, J.Q., Rana, T.M., Zhou, R., 2016. miR-34 modulates innate immunity and ecdysone signaling in *Drosophila*. PLoS Pathog. 12, e1006034.

Xu, L., Ling, Y., Wang, Y., Wang, Z., Hu, B., Zhou, Z., Hu, F., He, K., 2015. Identification of differentially expressed microRNAs between *Bacillus thuringiensis* Cry1Ab-resistant and -susceptible strains of *Ostrinia furnacalis*. Sci. Rep. 5, 15461.

Ye, X., Liu, Q., 2008. Expression, purification, and analysis of recombinant Drosophila dicer-1 and dicer-2 enzymes. Methods Mol. Biol. 442, 11–27.

Yu, T., Li, X., Coates, B.S., Zhang, Q., Siegfried, B.D., Zhou, X., 2018. microRNA profiling between *Bacillus thuringiensis* Cry1Ab-susceptible and -resistant European corn borer, *Ostrinia nubilalis* (Hubner). Insect Mol. Biol. 27, 279–294.

Zardo, G., Ciolfi, A., Vian, L., Billi, M., Racanicchi, S., Grignani, F., Nervi, C., 2012. Transcriptional targeting by microRNA-polycomb complexes: a novel route in cell fate determination. Cell Cycle 11, 3543–3549.

Zhang, G., Hussain, M., O'neill, S.L., Asgari, S., 2013. *Wolbachia* uses a host microRNA to regulate transcripts of a methyltransferase contributing to dengue virus inhibition in *Aedes aegypti*. Proc. Natl. Acad. Sci. U. S. A. 110, 10276–10281.

Zhang, G., Hussain, M., Asgari, S., 2014a. Regulation of arginine methyltransferase 3 by a *Wolbachia*-induced microRNA in *Aedes aegypti* and its effect on *Wolbachia* and dengue virus replication. Insect Biochem. Mol. Biol. 53, 81–88.

Zhang, J., He, Q., Zhang, C.D., Chen, X.Y., Chen, X.M., Dong, Z.Q., Li, N., Kuang, X.X., Cao, M.Y., Lu, C., Pan, M.H., 2014b. Inhibition of BmNPV replication in silkworm cells using inducible and regulated artificial microRNA precursors targeting the essential viral gene lef-11. Antiviral Res. 104, 143–152.

Zhang, X., Zheng, Y., Jagadeeswaran, G., Ren, R., Sunkar, R., Jiang, H., 2014c. Identification of conserved and novel microRNAs in *Manduca sexta* and their possible roles in the expression regulation of immunity-related genes. Insect Biochem. Mol. Biol. 47, 12–22.

Zhang, X., Aksoy, E., Girke, T., Raikhel, A.S., Karginov, F.V., 2017. Transcriptome-wide microRNA and target dynamics in the fat body during the gonadotrophic cycle of *Aedes aegypti*. Proc. Natl. Acad. Sci. U. S. A. 114, E1895–E1903.

Zhou, Q., Wang, Z., Zhang, J., Meng, H., Huang, B., 2012. Genome-wide identification and profiling of microRNA-like RNAs from *Metarhizium anisopliae* during development. Fungal Biol. 116, 1156–1162.

Zhou, Y., Liu, Y., Yan, H., Li, Y., Zhang, H., Xu, J., Puthiyakunnon, S., Chen, X., 2014. miR-281, an abundant midgut-specific miRNA of the vector mosquito *Aedes albopictus* enhances dengue virus replication. Parasit. Vectors 7, 488.

Zhu, M., Wang, J., Deng, R., Xiong, P., Liang, H., Wang, X., 2013. A microRNA encoded by *Autographa californica* nucleopolyhedrovirus regulates expression of viral gene ODV-E25. J. Virol. 87, 13029–13034.

Zhu, M., Wang, J., Deng, R., Wang, X., 2016. Functional regulation of an Autographa californica nucleopolyhedrovirus-encoded microRNA, AcMNPV-miR-1, in baculovirus replication. J. Virol. 90, 6526–6537.

Zug, R., Hammerstein, P., 2012. Still a host of hosts for *Wolbachia*: analysis of recent data suggests that 40% of terrestrial arthropod species are infected. PLoS One 7, e38544.

CHAPTER THREE

The Honeybee Queen: The Implications of Eusociality on Parasite-Mediated Competition

Ivan Meeus, Guy Smagghe
Department of Plants and Crops, Faculty of Bioscience Engineering, Ghent University, Ghent, Belgium

Contents

Abstract

A reliable pollination of crops is provided by integrated crop pollination (ICP). It combines the use of managed and wild pollinators in a location-specific manner. The honeybee is our best known ally and an essential part of ICP. Here we highlight two main aspects of the honeybee's biology and use some anthropomorphism elements to explain our point of view. The first is on social behaviour, the shift from single mating towards a promiscuous lifestyle was essential to develop full eusociality. This also means that some half sisters are stuck in a social framework and raise fertile offspring, while having low kinship with them. Second, a queen develops a large workforce of bees, with a division of labour. Aside from nursing and foraging, also time is allocated towards defence. We question if more assertive strategies are present. We speculate on a workforce carrying parasites, to perform parasite-mediated competition with other, yet solitary, bees. We discuss two mechanisms: (i) where social immunity tempers the effect of a specific parasite in a social bee; (ii) current domestication practises could also facilitate parasite-mediated competition in both directions.

Advances in Insect Physiology, Volume 55
ISSN 0065-2806
https://doi.org/10.1016/bs.aiip.2018.07.006

1. THE HONEYBEE: AN IMPORTANT MANAGED POLLINATOR

Current agricultural practises have increased the demand for animal pollination (Aizen and Harder, 2009). Particularly, fruit, vegetable and nut crops require cross pollination and provide essential (micro)nutrients to ensure a rich diet (Chaplin-Kramer et al., 2014). The concept of integrated crop pollination (ICP) is installed in order to support pollinators and provide reliable and economical pollination of crops. It combines the use of managed pollinators in synergy with wild pollinator habitat management (Isaacs et al., 2017). Such practices are crucial, the worldwide economic value of pollination is amounted to €153 billion (Gallai et al., 2009). An important species in ICP is the honeybee, allowing the allocation of 10 thousands of bees to a target environment by the simple measure of relocating a beehive.

Aside from its importance in agriculture, the honeybee has a fascinating biology. Its social structure with altruistic infertile workers challenged the earliest ideas on species evolution. Here we discuss two strategies of how honeybee queens could improve their fitness. We formulate some hypotheses, which can direct future study design and research on parasite interactions between social and solitary bee species. Hereto we first need to understand the implications of eusocial behaviour.

2. EUSOCIAL *APIS MELLIFERA* DOES NOT SERVE ALL WORKER BEES

George Orwell's modern fable "Animal Farm" depicts the stepwise rise of a dictatorship. "The Seven Commandments" of Animalism, initially serving all, eroded. Perhaps the seventh was the most striking: "All animals are equal"; it was eventually prolonged with "but some are more equal than others". Here we argue that the above has some analogy with the evolution of social behaviour of honeybees: "All sister alleles are equal but some are more equal than others". But let us explain; kin theory tells why sociality of worker bees (infertile workforce serving the queen's offspring) could evolve. Basically, this is because of the haplodiploid-mating system, the worker bees are genetically more related with their full sisters (relatedness, $r_{mean} = 0.75$) than with their own offspring ($r_{mean} = 0.5$). Serving their queen and raising up full sisters is genetically more rewarding compared with diploid-mating organisms. In analogy with the initials of *Animalism*, "all sister alleles are equal". But did the honeybee queen hold her end of the bargain? The above presented numbers only stand if the queen mates only once; indeed all eusocial hymenopterans evolved from monogamous ancestors (Hughes et al., 2008b). Once the work division was created and the workers lost their fertility for the greater good of their shared alleles, the queen can shift away from this strategy (Boomsma et al., 2011). Hence some altruistic workers are stuck in a social framework, while there is low kinship, with lower survival potential of their own unique alleles. The r_{mean} of 0.75 with full sisters drops to an r_{mean} of 0.25 for half sisters. One could argue that at least a fraction of the workers genes are high-jacked by the queen; to complete the allegory "all sister alleles are equal but some are more equal than others".

In contrast to the individual rights of some half sisters, multiple mating or polyandry comes with benefits for the queens alleles. Polyandry is omnipresent in eusocial insects and has evolved within ants, bees and wasps (Crozier and Page, 1985). Also within the bees, the monogamous species, like most bumble bees species, have smaller nests and have less strict separation of division of labour (Huth-Schwarz et al., 2011). Worker bees dedicate themselves towards foraging, nursing, defence, nest care without the bourdon of sexual reproduction. Here we argue that aside from activities directly needed for nest development and defence (Breed et al., 1990), a task division can evolve which are linked with direct competitive interactions with other pollinators. In defence, a worker honeybee is willing to pay the ultimate price, as it dies after stinging; yet it is unknown if worker bees are allocated to a more assertive or proactive strategies. Here we introduce the concept of parasite-mediated competition, and its implication in (eu) social insects.

3. COMPETITION MEDIATED BY PARASITES

Multihost parasites can be considered as potential powerful weapons, used to influence coexistence or exclusion of (originally) sympatric species (Hudson and Greenman, 1998; Strauss et al., 2012). Hosts have differences in tolerance towards a parasite and different associated immune costs, which results in the existence parasite-mediated competition (Fig. 1A). Examples

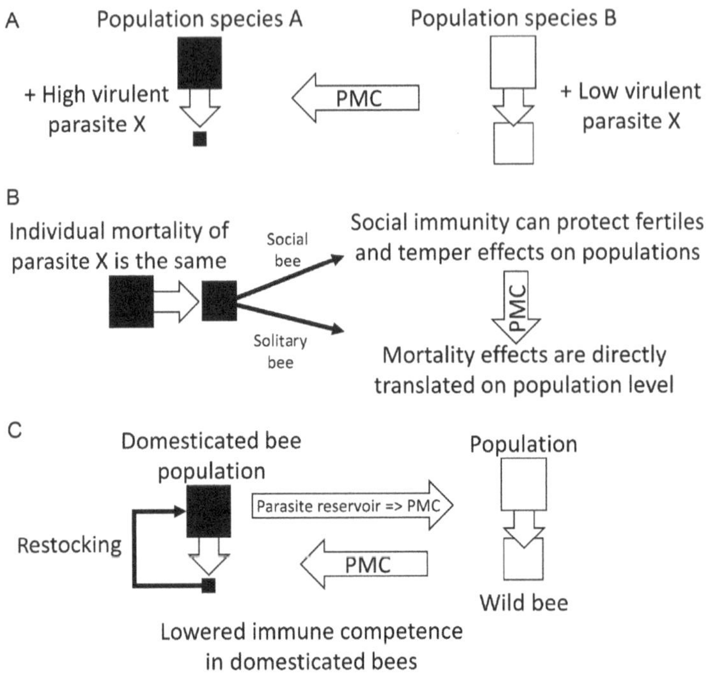

Fig. 1 Parasite-mediated competition (PMC) between different bee species. (A) We give an example of parasite-mediated competition; species B has a higher tolerance towards a specific parasite X; this gives a competitive advantage of species B over A. The graph represents a snapshot in time, we are aware that parasite reproduction correlates the amount of uninfected individuals (Anderson and May, 1982), and this interaction is a dynamic process. (B) We give the hypothetical example for a parasite with a same mortality in two species, a solitary and social species. In the case that such a parasite is "controlled" by social immunity, then the impact on the fitness (the number of offspring) can be tempered by preventing infection of the queen. A dead solitary species means the end of offspring reproduction. (C) A parasite-mediated competition can arise when bee domestication results in less immunocompetent bees. If the domesticated has a lower tolerance to the parasite this could lead to a competitive advantage of the wild solitary bee. Yet, domesticated bees are restocked, and gained habitat for the solitary bee will therefore again be refilled. This restocking can be problematic as it can result in a pathogen reservoir, with ideal parasite reproduction as there is a constant influx of new uninfected domesticated bees.

of parasite-mediated competition are mainly found when studying species invasions, with, for example, a competitive advantage of the invasive host introducing new parasites (or new variants) causing infections in native, phylogenetic related, species (Blackburn and Ewen, 2017). Also pathogens present in low prevalence could be important, for instance in the decline of red squirrels in the United Kingdom (Tompkins et al., 2003).

4. PARASITE-MEDIATED COMPETITION AND SOCIAL BEES

For eusocial and social bees no examples of parasite-mediated competition are described, yet parasite screening in pollinators revealed that multihost dynamics are present and thus parasite-mediated competition could be present. We argue that within social species this mechanism can be further optimised, by allocating infected workers towards foraging. Indeed in social bees a hierarchic structure exists, with different contact networks between nest members, and therefore different likelihoods to get infected (Schmid-Hempel, 2017). Herein social immunity is important, a parasite can be prevented to reach the fertile founding queen and/or offspring; parasite infection throughout the whole colony can be prevented by a division of labour (Schmid-Hempel and Schmid-Hempel, 1993). This provokes the hypothesis that parasites, which are "controlled" by social immunity, are ideal candidates for parasite-mediated competition between a social species (partially protect by social immunity) and solitary species (Fig. 1B).

Up to date we have too few insight in virulence dynamics of parasites and viruses in different bee species to conclude that parasite-mediated competition with an infected workforce exists. It can be argued that parasites in social species have higher likelihood to develop a lower virulence (Hughes et al., 2008a), based on the virulence-transmission trade-offs (Rigaud et al., 2010), and a high means of vertical transmission (parent to offspring). Yet such predictions are not without controversy, and virulence evolution within multihost systems is notoriously complex (Rigaud et al., 2010; Schmid-Hempel, 2017). Yet different virulence in species can be expected and can even be quite extreme; if the alternate host is not needed for parasite transmission it can lead to highly lethal infections (Leggett et al., 2013).

5. COMPETITION MEDIATED BY PARASITES AND DOMESTICATED BEES

Human interest in honeybees and keeping them in bee hives can be traced back already 3000 years (Bloch et al., 2010). Especially in modern times the selection process towards high honey yield and ease to handle got intensified. Nowadays, multiple stressors impacting honeybee health are present, and beekeepers suffer severe losses of hives. Yet the honeybee has us as a powerful ally. Locally collapsed hives can be restocked, which is an understandable practise from the beekeepers point of view. Yet on the longer term this strategy can be counterproductive. The domesticated honeybee seems to be less immune competent when compared to feral honeybees (Lopez-Uribe et al., 2017; Youngsteadt et al., 2015). This is not only troublesome for the honeybee itself, as less immune competent domesticated bees could also act as pathogen reservoir for local bee species. Such mechanism could harm local wild bee populations (Meeus et al., 2011). We know that many pathogens can infect both managed honeybees and wild bumble bees (Fürst et al., 2014; McMahon et al., 2015; Ravoet et al., 2014), predisposing them to potential parasite-mediated competition, possible in both directions (Fig. 1C).

6. RESEARCH DIRECTIONS

When one wants to study parasite-mediated competition we recommend to follow-up infectivity of parasites in different hosts, using isolates obtained from different hosts. Aside from infectivity, parasite load should also be determined. Here an important challenge is the development of good proxy measures to determine the actual damage caused, in terms of fitness loss, by the parasite. Distinguishing parasite tolerance and resistance is also important, as the second will be more important in terms of parasites clearances, while the former can stimulate parasite transmission. A tolerant host developed strategies to reduce the negative impact of an infection on host fitness (Medzhitov et al., 2012). Resistance is an immune function, eradicating the infection, balancing the gain of parasite clearance with the level of immunopathology, i.e., the impact on host fitness caused by the immune response (Casadevall and Pirofski, 1999). Understanding the role of social behaviour in relation to virulence tolerance is an exciting area for future research. Finally, the virulence of a specific parasite is determined by the environment of the host. With multiple inference mechanism present between the drivers of bee decline, all influencing host parasite dynamics (Meeus et al., 2018).

ACKNOWLEDGEMENTS

We acknowledge the Foundation Research-Flanders (FWO) and the Belgian Science Policy Office (Belspo).

CONFLICTS OF INTEREST

None.

REFERENCES

Aizen, M.A., Harder, L.D., 2009. The global stock of domesticated honey bees is growing slower than agricultural demand for pollination. Curr. Biol. 19 (11), 915–918.

Anderson, R.M., May, R.M., 1982. Coevolution of hosts and parasites. Parasitology 85 (2), 411–426.

Blackburn, T.M., Ewen, J.G., 2017. Parasites as drivers and passengers of human-mediated biological invasions. EcoHealth 14 (1), S61–S73.

Bloch, G., Francoy, T.M., Wachtel, I., Panitz-Cohen, N., Fuchs, S., Mazar, A., 2010. Industrial apiculture in the Jordan valley during Biblical times with Anatolian honeybees. Proc. Natl. Acad. Sci. U.S.A. 107 (25), 11240–11244.

Boomsma, J.J., Beekman, M., Cornwallis, C.K., Griffin, A.S., Holman, L., Hughes, W.O.H., Keller, L., Oldroyd, B.P., Ratnieks, F.L.W., 2011. Only full-sibling families evolved eusociality. Nature 471 (7339), E4–E5.

Breed, M.D., Robinson, G.E., Page, R.E., 1990. Division of labor during honey bee colony defense. Behav. Ecol. Sociobiol. 27 (6), 395–401.

Casadevall, A., Pirofski, L.A., 1999. Host-pathogen interactions: redefining the basic concepts of virulence and pathogenicity. Infect. Immun. 67 (8), 3703–3713.

Chaplin-Kramer, R., Dombeck, E., Gerber, J., Knuth, K.A., Mueller, N.D., Mueller, M., Ziv, G., Klein, A.M., 2014. Global malnutrition overlaps with pollinator-dependent micronutrient production. Proc. R. Soc. B Biol. Sci. 281 (1794). https://doi.org/10.1098/rspb.2014.1799.

Crozier, R.H., Page, R.E., 1985. On being the right size—male contributions and multiple mating in social Hymenoptera. Behav. Ecol. Sociobiol. 18 (2), 105–115.

Fürst, M.A., McMahon, D.P., Osborne, J.L., Paxton, R.J., Brown, M.J.F., 2014. Disease associations between honeybees and bumblebees as a threat to wild pollinators. Nature 506 (7488), 364–366.

Gallai, N., Salles, J.M., Settele, J., Vaissiere, B.E., 2009. Economic valuation of the vulnerability of world agriculture confronted with pollinator decline. Ecol. Econ. 68 (3), 810–821.

Hudson, P., Greenman, J., 1998. Competition mediated by parasites: biological and theoretical progress. Trends Ecol. Evol. 13 (10), 387–390.

Hughes, D.P., Pierce, N.E., Boomsma, J.J., 2008a. Social insect symbionts: evolution in homeostatic fortresses. Trends Ecol. Evol. 23 (12), 672–677.

Hughes, W.O.H., Oldroyd, B.P., Beekman, M., Ratnieks, F.L.W., 2008b. Ancestral monogamy shows kin selection is key to the evolution of eusociality. Science 320 (5880), 1213–1216.

Huth-Schwarz, A., Leon, A., Vandame, R., Moritz, R.F.A., Kraus, F.B., 2011. Mating frequency and genetic colony structure of the neotropical bumblebee Bombus wilmattae (Hymenoptera: Apidae). Apidologie 42 (4), 519–525.

Isaacs, R., Williams, N., Ellis, J., Pitts-Singer, T.L., Bommarco, R., Vaughan, M., 2017. Integrated crop pollination: combining strategies to ensure stable and sustainable yields of pollination-dependent crops. Basic Appl. Ecol. 22, 44–60.

Leggett, H.C., Buckling, A., Long, G.H., Boots, M., 2013. Generalism and the evolution of parasite virulence. Trends Ecol. Evol. 28 (10), 592–596.

Lopez-Uribe, M.M., Appler, R.H., Youngsteadt, E., Dunn, R.R., Frank, S.D., Tarpy, D.R., 2017. Higher immunocompetence is associated with higher genetic diversity in feral honey bee colonies (Apis mellifera). Conserv. Genet. 18 (3), 659–666.

McMahon, D.P., Furst, M.A., Caspar, J., Theodorou, P., Brown, M.J.F., Paxton, R.J., 2015. A sting in the spit: widespread cross-infection of multiple RNA viruses across wild and managed bees. J. Anim. Ecol. 84 (3), 615–624.
Medzhitov, R., Schneider, D.S., Soares, M.P., 2012. Disease tolerance as a defense strategy. Science 335 (6071), 936–941.
Meeus, I., Brown, M.J.F., De Graaf, D.C., Smagghe, G., 2011. Effects of invasive parasites on bumble bee declines. Conserv. Biol. 25 (4), 662–671.
Meeus, I., Pisman, M., Smagghe, G., Piot, N., 2018. Interaction effects of different drivers of wild bee decline and their influence on host-pathogen dynamics. Curr. Opin. Insect Sci. 26, 136–141.
Ravoet, J., De Smet, L., Meeus, I., Smagghe, G., Wenseleers, T., de Graaf, D.C., 2014. Widespread occurrence of honey bee pathogens in solitary bees. J. Invertebr. Pathol. 122, 55–58.
Rigaud, T., Perrot-Minnot, M.J., Brown, M.J.F., 2010. Parasite and host assemblages: embracing the reality will improve our knowledge of parasite transmission and virulence. Proc. R. Soc. B Biol. Sci. 277 (1701), 3693–3702.
Schmid-Hempel, P., 2017. Parasites and their social hosts. Trends Parasitol. 33 (6), 453–462.
Schmid-Hempel, P., Schmid-Hempel, R., 1993. Transmission of a pathogen in *Bombus terrestris*, with a note on division of labor in social insects. Behav. Ecol. Sociobiol. 33 (5), 319–327.
Strauss, A., White, A., Boots, M., 2012. Invading with biological weapons: the importance of disease-mediated invasions. Funct. Ecol. 26 (6), 1249–1261.
Tompkins, D.M., White, A.R., Boots, M., 2003. Ecological replacement of native red squirrels by invasive greys driven by disease. Ecol. Lett. 6 (3), 189–196.
Youngsteadt, E., Appler, R.H., Lopez-Uribe, M.M., Tarpy, D.R., Frank, S.D., 2015. Urbanization increases pathogen pressure on feral and managed honey bees. PLoS One 10 (11). https://doi.org/10.1371/journal.pone.0142031.

CHAPTER FOUR

The Potential Control Strategies Based on the Interaction Between Indoor Cockroaches and Their Symbionts in China

Xian Cui Zhang, Fan Zhang[1]

Key Laboratory of Animal Resistance Research, College of Life Science, Shandong Normal University, Jinan, The People's Republic of China

[1]Corresponding author: e-mail address: zhangfan0531@163.com

Contents

Advances in Insect Physiology, Volume 55
ISSN 0065-2806
https://doi.org/10.1016/bs.aiip.2018.07.001

Abstract

Cockroaches, the important indoor sanitary pests in urban worldwide. In China, there are six species of dominant indoor cockroaches which are increasingly hard to control due to their high adaptability, strong fecundity and extensive insecticide resistance. These cockroaches harbour abundant microorganisms varying from bacteria to viruses, fungi and protists, which have formed an interdependent symbiotic relationship with hosts in the process of long-term coevolution. A growing research has revealed that symbionts can modulate limited metabolic networks, thereby expanding ecological niches of host insects as a significant regulatory factor in insect growth and development. Therefore, in-depth study on the mechanisms of symbiont–insect interactions can shed light on a new perspective for pest prevention and control technology. Here, we give a general overview on classification and hazards of indoor cockroaches in China, as well as species, distribution and transmission mechanism of symbionts within cockroaches. Besides, we mainly discuss the interactions between insects and symbionts from the following aspects: nutritional physiology, reproductive regulation, pesticide resistance, defence, behaviour and so on. Based on the earlier symbiotic interactions, we propose some promising new insights into integrated pest management of cockroaches (i.e., utilizing symbionts to control cockroaches), such as application of *Wolbachia* to restrict population number, utilization of interactions between symbionts and entomopathogenic fungi including synergism or antagonism to accelerate host death, limiting host's vectorial capacity via paratransgenesis, push–pull strategy.

1. CLASSIFICATION, HAZARDS AND CONTROL OF COMMON INDOOR COCKROACHES

1.1 Classification

At present, there are more than 5000 kinds of cockroaches around the world, and there are 252 kinds of cockroaches in China. Cockroach is divided into two types: home and wild. The wild species are the majority, and less than 1% of the species live indoors.

The species and distribution of indoor sanitary pest cockroach are widely studied in China. It is basically divided into five families: Blattidae, Blaberidae, Blattellidae, Corydiidae, Epilamparidae, 11 genera: *Periplaneta, Neostyloqyga, Hebardina, Blatta, Pyenoseelus, Blattella, Assablatta, Polyphaga, Eupolyphaga, Opisthoplatia, Rhabdblatta*, 21 species: *Periplaneta americana, Periplaneta fuliginosa, Periplaneta brnnea, Periplaneta fallax, Periplaneta australasiae, Periplaneta japonica, Neostyloqyga rhombifolia, Hebardina concinna, Blatta orientalis, Pyenoseelus surinanensis, Blattella germanica, Blattella liturieollis, Blattella latistriga, Assablatta keyotensis, Polyphaga plancyi, Eupolyphaga sinensis, Eupolyphaga limbata, Eupolyphaga yunnanensis, Eupolyphaga thibetana,*

Opisthoplatia orientalis, *Rhabdblatta sinuata* (Pan and Qiang, 2017). Among them, there are 6 species of cockroaches in Northwest China, 9 in North China, 7 in Northeast China, 15 in Southwest China, 12 in Central China, 15 in East China and 11 in South China (Table 1).

There are six species of cockroaches which are most harmful to human health and life: *B. germanica*, *P. americana*, *P. fuliginosa*, *P. brnnea*, *P. australasiae*, and *P. japonica* (Pan and Qiang, 2017). They are all nationally distributed and the vertical distribution decreases with the increase of altitude. Among them, *B. germanicar* are becoming the main dominant species all over the country due to the insect's rapid propagation, high adaptability and resistance to various chemical insecticides.

1.2 Hazards

Indoor sanitary pest cockroaches follow human living environment, bit human being, chew on clothing and supplies, and contaminate food, seriously disturbing human rest and life. They also can drill into modern communication equipment or high precision automatic control apparatus causing accidents.

A few cockroaches can harm crops, such as *P. surinamensis*, *P. americana*, *P. australasiae* and *Diploptera ditiseoides*. *P. surinamensis* is harmful to rice, sugarcane and spinach in southern China. *P. americana* and *P. australasiae* can endanger plant seedlings in greenhouse.

Indoor cockroach is one of the most important vectors, which can carry and spread pathogenic microorganism bacteria, fungi, intestinal parasite eggs and virus. They can transmit nearly 150 species of bacteria, 60 species of pathogenic fungi, 45 species of parasitic worms and 90 species of protozoa to human life, whether through biological or mechanical transmission. In addition, cockroaches can carry viruses such as acute hepatitis b virus, enterovirus, poliovirus, Newcastle disease virus and adenovirus (Goralska et al., 2013; Vazirianzadeh et al., 2014).

Cockroaches are the vital allergen of human skin, rhinitis, asthma, chronic idiopathic urticaria, etc. Currently, 12 groups of cockroach allergens are listed in the official Allergen Nomenclature database maintained by the World Health Organization/International Union of Immunological Societies (WHO/IUIS) Subcommittee (www.allergen.org). These groups comprise allergens with a wide variety of structures and functions (Pomes et al., 2017). Cockroach allergy is established as an important cause of asthma particularly among children and young adults living in innercity

Table 1 Classification and Distribution of Indoor Sanitary Pest Cockroach in China (Pan and Qiang, 2017)

Family	Genus	Species	Northwest China	North China	Northeast China	Southwest China	Central China	East China	South China
Blattidae	*Periplaneta*	*Periplaneta americana*	++	++	++	++	++	++	++
		Periplaneta fuliginosa	+	+	+	++	++	++	++
		Periplaneta brnnea	—	+	+	++	+	+	+
		Periplaneta fallax	—	—	—	+	—	+	—
		Periplaneta australasiae	+	+	—	+	+	+	+
		Periplaneta japonica	—	++	++	—	++	+	+
	Neostyloqyga	*Neostyloqyga rhombifolia*	—	—	—	+	—	+	+
	Hebardina	*Hebardina concinna*	—	—	—	+	—	—	—
	Blatta	*Blatta orientalis*	—	—	—	—	+	—	—
Blaberidae	*Pyenoseelus*	*Pyenoseelus surinanensis*	+	—	—	+	+	+	+
Blattellidae	*Blattella*	*Blattella germanica*	++	++	++	++	++	++	+++
		Blattella liturieollis	—	—	—	+	+	+	—
		Blattella latistriga	—	—	—	+	+	+	—

Corydiidae	*Assablatta*	*Assablatta keyotensis*	—	—	—	+	+	+	—
	Polyphaga	*Polyphaga plancyi*	—	+	+	—	—	+	—
	Eupolyphaga	*Eupolyphaga sinensis*	+	+	+	+	+	+	+
		Eupolyphaga limbata	—	—	—	+	—	—	—
		Eupolyphaga yunnanensis	—	—	—	+	—	—	+
		Eupolyphaga thibetana	—	—	—	+	—	—	—
Epilamparidae	*Opisthoplatia*	*Opisthoplatia orientalis*	—	+	—	—	—	+	+
	Rhabdblatta	*Rhabdblatta sinuata*	—	—	—	—	—	—	—

Note: + represents distribution; ++ and +++ represent a large population; — represents no distribution.

environments, put forward a challenge for disease control (Pomes et al., 2017). *B. germanica* and *P. americana* are the main focus of research on allergies, they are dominant in temperate and tropical regions, respectively, among them, *B. germanica* prefers cool and dry climate, whereas *P. americana* is common in hot and humid environment (Pomés, 2008; Pomés and Arruda, 2014). In the United States, the prevalence of cockroach allergy ranges from 17% to 41% and the cockroach allergens were detected in 85% of homes in innercity. Based on the skin prick testing, 60%–80% of innercity children with asthma were sensitive to cockroaches (Do et al., 2016). In large- and medium-sized cities of China, the incidence of cockroach allergy among children with asthma is up to 28% (Ma et al., 2011). In south China, cockroach is also the major cross-reactive allergen in shrimp-sensitized rural children (Yang et al., 2018).

Endotoxin is a kind of lipid polysaccharide which is naturally occurring in the outermembrane of Gram-negative bacteria and is a ubiquitous indoor biocontaminant that causes the development and severity of asthma and other respiratory symptoms (Kulhankova et al., 2012; Thorne et al., 2009). Cockroach infestation is one of the sources of indoor endotoxins but the roles of cockroaches and contamination mechanisms are unclear. The faeces of the cockroaches may contain allergens and endotoxins from the digestive tract (Tungtrongchitr et al., 2012) to form inhalable particles through which inhalation enters the body. Some studies suggest that the bacterial load in the cockroach's gut is related to the bacterial load in the environment (Fu et al., 2009; Rivault et al., 1993), which means that the levels of endotoxin in faeces may be related to the environmental bacterial load. Wild cockroaches contain nearly 10 times more endotoxins in their body than reared cockroaches, for example, the faeces of *P. australasiae* can have a much higher level of endotoxins than that in the faeces of *Supella longipalpa*, and *B. germanica* (Lai, 2017). Studies showed that endotoxins regulate the allergic reactions to cockroach allergens and the coadministration of both cockroach allergens and endotoxins increases pulmonary and systemic responses (Kulhankova et al., 2009; Natarajan et al., 2011; Li et al., 2012).

1.3 Pesticide Resistance and Biological Control

Resistance is a genetically based characteristic that allows insects to tolerate concentrations of insecticides that are lethal to other individuals of the same species (Tabashnik et al., 2014). The proliferation of resistant individuals makes it more difficult to control the insects and subsequently damage caused by the pest. Sometimes, resistance leads to incorrect practices like increasing the frequency or quantity of insecticide treatments.

The commonly used pesticides for cockroaches control are pyrethroid, carbamate, organochlorine and organophosphorus. Many cockroaches are remarkably resistant to chemical pesticides. Take the German cockroach, for example, for 42 active ingredients in different parts of the world, *B. germanica* is seventh in the "Top 20" chart of the most insecticide-resistant insect species (Zhang et al., 2014).

Resistance to insecticides in field populations of *B. germanica* was first detected at Corpus Christi, Texas, United States in 1953, where cockroach's resistance to chlordane, lindane and DDT was recorded (Heal et al., 1953). Subsequently, it was successively discovered that *B. germanica* produced different levels of resistance to organochlorine, organophosphorus, carbamate, pyrethroid classes of insecticides. By 2018, numerous studies have documented widespread insecticide resistance of the *B. germanica* (Wu and Appel, 2017). Cockroaches show varying degrees of pesticide resistance to various types of pesticides, including low resistance, moderate resistance and high resistance. In field populations of the *B. germanica* from Singapore, resistance level reported till now is about 62.50 × against the carbamates (propoxur), up to 28.80 × against organophosphorus (chlorpyrifos), up to 468.00 × against pyrethroids (beta-cyfluthrin and deltamethrin), and 10.0 × against phenyl pyrazole (fipronil). However, moderate level of resistance is found in cockroaches against neonicotinoids, i.e., imidacloprid (0.8–3.8 ×) and the oxadiazines like indoxacarb (1.4–5.3 ×) (Naqqash et al., 2016). Other species of cockroaches are also resistant to insecticides. For example, American cockroach and smokybrown cockroach populations were found to be resistant to chlorinated hydrocarbon insecticides during the period of time when these insecticides were used extensively for cockroach control. In addition, resistance in American cockroach to trichlorfon in China and Brown cockroach to diazinon in the United States have been reported (Pardo et al., 2017).

With the development of resistance, biological control has become a promising strategy. Biological control is a method of using natural enemies, pathogenic microorganisms (viruses, pathogenic bacteria and pathogenic fungi) and their metabolites to control pests.

In nature, there are many kind of natural enemies of the cockroach, mainly including parasitoids and predators. There are about 15 kinds of parasitoid natural enemies such as parasitic wasps, parasitic mite, etc.

Predatory natural enemies have 18 kinds, such as spiders, centipedes, scorpions, etc. Predatory natural enemies act on the central nervous system of cockroach by releasing venom, disrupting the normal behaviour of cockroach, causing the cockroach to paralyse and then prey. Adults and nymphs of cockroach are its control objects. Parasitoid natural enemies such as parasitoid wasps can produce eggs in the egg pods of cockroaches, after hatching, the young bees absorb nutrients from the egg pods of cockroaches in order to damage the incubation of cockroaches. For example, *Ampulex compressa* relies on cockroaches such as *P. americana* for food for its offspring. The *A. compressa* female attacks a cockroach, stings it and drags the docile host to a hiding place, where it glues an egg to the coxa of one of the mesothoracic legs of the cockroach. When the wasp's egg hatches, the larva first imbibes haemolymph through a hole in the thoracic cuticle of the still living cockroach. About 7 days after oviposition (at 27°C), the larva moves inside the cockroach and feeds on the interior organs, causing the death of the host (Herzner et al., 2013). Parasitoid wasps stung the cockroach in the thoracic and head subesophageal ganglion and delivered a cocktail of neurotoxins directly inside the prey's cerebral ganglia, causing the cockroach to lose its ability to move freely and become a wasp parasitic or prey (Gal and Libersat, 2010). Besides, venom proteins from parasitoid wasps damaged the host immune system and disrupted the host insect reproductive and endocrine system (Moreau and Asgari, 2015). Furthermore, the parasitic wasp was an important vector insect for the transmission of pathogenic microorganisms, predation or parasitism between parasitic wasp and target insect can be available to transmitted fungi, viruses and other pathogenic microorganisms, infected or preyed harmful organisms, thereby achieved a lasting insecticide effect (Burke and Strand, 2012). However, vector cockroach's damage mainly occurs indoors, which greatly limits the practical application of parasitic natural enemies.

Periplaneta fuliginosa Densovirus (PfDNV) has been successfully used in cockroach control. PfDNV was isolated from the *P. fuliginosa* in Wuhan in China, which can infect and lead to the death of *P. fuliginosa* and *B. germanica* (Jiang, 2003). DNV entered the cockroach cell by adsorption of structural proteins and then inhibited the S-phase replication of the host cell nucleus and activated its nucleic acid replication. The mass propagation of DNV brought about the nuclear membrane rupture, caused the whole tissue cell lesion, especially infected the cockroach fat body, thereby destroyed the host energy storage system, eventually created the host action ability weakened and gradually led the host to die (Liang et al., 2010;

Liu et al., 2012). At present, cockroach bait made by PfDNV has been put into production, in which Wuda Dulida is the most successful.

Insect pathogenic bacteria due to its high specificity of the insecticidal activity and harmless to nontarget insect has been widely used as the biological insecticide. Nishiwaki purified the *Bacillus sphaericus* A3–2 and *Bacillus cereus* which had insecticidal activity to the *B. germanica* from the regurgitated fluid of *Myrmeleon bore* larvae. As for *B. thuringiensis* (Bt), which is widely used in the control of lepidoptera insects, insecticidal crystal protein only can play a insecticidal role in the insect alkaline midgut, it may not perform function in the partial acid gut environment of the cockroaches, so it does not have higher application value in the control of cockroaches (Marroquin et al., 2000).

The pathogenic fungi also have been successfully used in the control of cockroaches. The pathogenic fungi of cockroaches are mainly *Metarhizium anisopliae* and *Beauveria bassiana* (Zhang et al., 2013). Entomopathogenic fungi invaded roaches via the host exoskeleton or cuticle and produced appressoria during the course of infection. Under suitable conditions, the germ tubes were germinated and formed. It reproduced quickly in vivo by penetrating the cuticle, and destroyed the host important structure such as fat body, and secreted a series of toxins such as cyclic peptide toxins, eventually led to death of the roaches (Gao and Yao, 2012). In addition to cuticle infection, entomopathogenic fungi occasionally made a digestive tract infection. Lopes and Alves found that baits amended with *M. anisopliae* conidia feeding cockroaches would cause the death of *B. germanica*, but the mortality rate was inferior to 30%, which may be related to the less residual conidia in the digestive tract and antimicrobial peptides produced by symbiotic bacteria. In addition, other fungi such as *Isaria fumosoroseus, Hirsutella thompsonaii* fungal infections initiate oxidative stress in the midgut, fat body, whole body and haemolymph of cockroach thereby suggesting them to be the target organs for oxidative damage, and the mortality rate can reach 80%, which indicated that fungi had high potentials for biocontrol agents against roaches.

As a new biological insecticide, entomopathogenic nematode has the characteristics of both insect natural enemy and pathogenic microorganism. It actively looked for the host and simultaneous released the symbiotic bacteria when it infected the host body cavity, making the host died from haematosepsis (Liu and Xia, 2005). *Steinernema carpocapsae* had a better control effect on cockroaches. When infected *P. americana, P. fuliginosa, B. orientalis* and *B. germanica* by the filter paper soaked in 1 mL water suspension containing 500,000 nematodes, the LT_{50} was 4.13 days for *B. germanica* and the insecticidal lethal time was shorter (Koehler et al., 1992). Besides, the highest population density of nematodes, 300,000 infective juveniles per 50 cm^3 sand, killed 93% of the *P. brunnea* and produced a more rapid mortality when compared with other population densities (Corpus and Sikorowski, 1992).

In addition, the control of cockroaches can be carried out through the genetic modification of the symbiotic. The genetic manipulation of symbiotic microorganisms can be divided into three types: heterologous association (Fig. 1A), genetic modification of microbial partners (paratransgenesis) (Fig. 1B) and disruption (dashed arrow) of the microbial partner (Fig. 1C) (Arora and Douglas, 2017). Heterologous associations and paratransgenesis will be discussed in detail in Section 6. The goal of disrupting microbial partners is to control insect pests by eliminating the

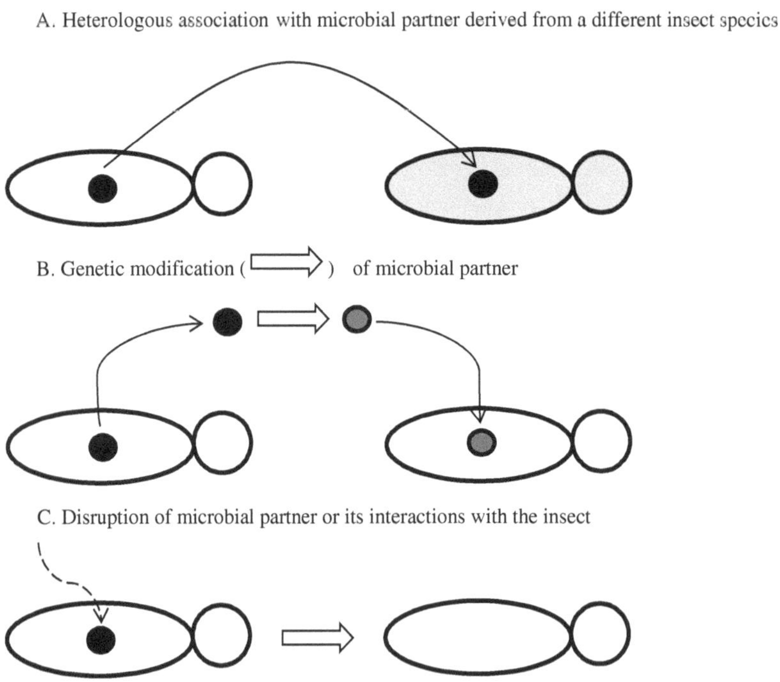

Fig. 1 Strategies for the insect pest control based on the manipulation of benign or beneficial microbial partners, resulting in either suppression of pest traits or impaired insect performance. (A) Heterologous associations. (B) Genetic modification of microbial partner. (C) Disruption (*dashed arrow*) of the microbial partner. The *black circle* represents a microbial taxon, with *dark grey* as genetically modified in (B). The insects are *white*, with *light grey* for a second insect species in (A) (Arora and Douglas, 2017).

microorganisms required for sustained insect growth, reproduction and survival, also can elimination the vertically transmission and horizontally transmission of obligate microbial partners.

2. THE SPECIES AND DISTRIBUTION OF SYMBIONTS

The symbiotic microorganisms are widely distributed in various parts of the insect. According to the different types of insects and microorganisms, the symbiotic microorganisms have different distribution. And it can be divided into two types: ectosymbionts and endosymbionts.

Endosymbionts are the microorganism that lives within the cells of an insect (Wang and Qu, 2017), mainly including yeast and eubacteria and proteobacteria, chlamydia and bacteroidetes also have distribution. The most common feature of the endosymbionts is genome shrinkage by gene lose, which reflecting the rich environment provided by the host insects, on the other hand, it also means that endosymbionts cannot live independently outside the host insects. These endosymbiont bacteria are usually found in the fat bodies of insects, gut epithelium, malpighian tubules, blood cells and ovarioles. The endosymbionts of insects are often divided into primary endosymbionts and secondary endosymbionts. In general, primary endosymbionts are necessary for insect hosts, and secondary endosymbionts are not. Primary endosymbionts often spread by insects matrilineal transmission, a vertical transmission between the generation of insects, like *Blattabacterium cuenoti* passed from mother to offspring (Tokuda et al., 2013). The secondary endosymbiotic bacteria can not only transmit vertically but also transmit horizontally among different host individuals. Compared to the secondary endosymbionts, the primary endosymbiotic bacteria are more closely related to the host. They can evolve with the host, and it is necessary for survival and reproduction of the host. *Buchnera aphidicola* in aphids is a typical example of primary endosymbiont. There are experimental evidence that *B. aphidicola* plays an important role in providing the host with essential amino acids and other nutrients (Gündüz and Douglas, 2009). Secondary endosymbionts may also be related to the adaptability or competitiveness of their hosts. For example, in some aphid individuals, *Candidatus Hamiltonella* can assist the host against predators and increase host heat tolerance (Dale and Moran, 2006; Rao et al., 2012). From the host's point of view, secondary endosymbionts are often facultative, with a short history of coevolution with a single species (Dale and Moran, 2006), and the secondary endosymbionts in the host are not stable

(Gueguen et al., 2010; Simon et al., 2003). In addition, secondary endosymbionts cannot be hosted in host-specific tissues and may infect extracellular tissues, such as blood cavities, fat, muscle tissue, nerve tissue or intestines, and newborns. Compared to primary endosymbionts, the occurrence of secondary endosymbionts is very low (Nancy et al., 2008).

A wide variety of endosymbionts also inhabit in cockroaches. Currently, detailed information on several endosymbionts of cockroaches is given in Table 2. But we have known very limited about the species of endosymbionts inhabited in the cockroaches. The endosymbiotic bacteria of the cockroach *B. cuenoti* was discovered in 1887, and it is also found in the weevil (Yang et al., 2014). *B. cuenoti* distributed in a special cell of fat body called cell bacteriocyte. Subsequently, *Blattabacterium* strain Bge, the primary endosymbiont of the omnivorous German cockroach, were identified and classified as a genus *Blattabacterium*, belonging to the class Flavobacteria in the phylum Bacteroidetes. They live in specialized cells in the host's abdominal fat body (López-Sánchez et al., 2009). *Blattabacterium* strain Bge codes for the complete urea cycle that, in combination with urease, produces ammonia as an end product.

Another typical endosymbionts is *Wolbachia*. Studies have shown that about 66% of insects are infected with *Wolbachia*, one of the widely known endosymbiotic bacteria known to date, is widely distributed in many species of the insect. *Wolbachia* also is an important endosymbiont in cockroaches, and always distributed in the head, chest, abdomen, salivary glands, digestive tract and other nonreproductive tissues and reproductive tissues. Its distribution is affected by many factors, among which the embryonic development process is the most important factor (Ros et al., 2009).

Ectosymbionts are the microorganisms that live outside the insect cells, which are concentrated in the digestive tract in the form of intestinal flora, including the bacteria that attach to the intestinal wall cells of insects and free in the intestinal cavity. The composition of insect intestinal flora is complex, including bacteria, fungi, archaea, viruses and protists. Among them, the species of bacteria is the most abundant and is also the largest number of insect intestinal microorganisms (Wang and Qu, 2017). Protists are mostly related to termites and wood roaches, and spread through social relations (Hongoh, 2010). Methanogenic archaea and fungi are often found in the intestinal tract of insects feed on wood or detritus, assisting the host to digest, such as termites and beetles (Engel and Moran, 2013). Insects can provide a stable living environment and the necessary nutrients for gut microbes, and the symbiotic microbes also participate in a variety of metabolic processes of insects, and the role of intestinal flora is particularly important.

Table 2 The Distribution and Function of Endosymbionts in Cockroaches

Insect	Bacteria	Category	Distribution	Function	Gram-Positive or Gram-Negative Bacteria
Blattella germanica	*Blattabacterium*	Flavobacterials	A special cell of fat body	Participate in nitrogen assimilation, uric acid degradation and nutrient provisioning (Sabree et al., 2009)	Gram-negative bacteria
	Wolbachia	Proteobacteria	The reproductive tract and the digestive tract of the head, thorax, abdomen, salivary gland, etc.	CI (cytoplasmic incompatibility), PI (induced parthenogenesis) and feminization, affecting insect reproductive via CI mechanisms, manipulating the host's reproductive (Gong et al., 2002; Ros et al., 2009; Shi et al., 2002)	Gram-negative bacteria
	Salmonella spp.	Proteobacteria	Gut tract	Assist the host defence drugs (Akinjogunla et al., 2012)	Gram-negative bacteria
	Bacteroides	Bacteroidetes	Gut tract	Carbohydrate metabolism and transport; assist the host defence entomopathogenic fungi (Backhed et al., 2005; Flint et al., 2008)	Gram-negative bacteria

Continued

Table 2 The Distribution and Function of Endosymbionts in Cockroaches—cont'd

Insect	Bacteria	Category	Distribution	Function	Gram-Positive or Gram-Negative Bacteria
	Lachnospira	Firmicutes	Gut tract	Hydrolyse polysaccharide, assist digestion, synthesize acetate, propionate and butyrate (Carrasco, 2014)	Gram-negative bacteria
	Pseudomonas	Proteobacteria	Gut tract	Secrete versatile secondary metabolites and protection from parasites and pathogens, to be resistant to both chloramphenicol and sulfamethoxazole (Akinjogunla et al., 2012; Palleroni, 1984; Zhang et al., 2013)	Gram-negative bacteria
	Acetobacteraceae	Proteobacteria	Gut tract	Participate in carbohydrate fermentation and lactate metabolism in flies (Thaochan et al., 2010)	Gram-negative bacteria
	Fusobacterium	Fusobacteria	Gut tract	Fermenting both glucose and amino acids (Carrasco, 2014)	Gram-negative bacteria

	Enterococcus sp.	Firmicutes	Gut tract	Antiphytopathogenic fungi activity (Oh et al., 2015)	Gram-positive bacteria
	Bacillus	Bacteriophyta	Gut tract	Inhibit microbial growth by secreting antifungal compounds and antibiotic like compounds (Martinari et al., 2002)	Gram-positive bacteria
	Weissella	Firmicutes	Gut tract	Produce many antimicrobial agents such as adhesion inhibitors, organic acids and bacteriocins (Stiles and Holzapfel, 1997)	Gram-positive bacteria
Blattellagermanica *Blatta orientalis*	*Rickettsia*	Proteobacteria	Digestive organs, salivary glands, reproductive organs, etc.	Participate in reproductive regulatory, increase host resistance and increase the host's ability to resist natural predators, high temperatures, or other lethal factors (Radek, 2000; Zhang et al., 2017)	Gram-negative bacteria
Periplaneta Americana *Blattellagermanica*	*Providencia*	Proteobacteria	Gut tract	Assist the host defence natural predators (Akinjogunla et al., 2012)	Gram-negative bacteria

The microbial flora in the intestinal tract of insects is extremely rich. There are both ectosymbionts located in the intestinal cavity and endosymbionts living in the intestinal epidermal cells. Including deformation (Proteobacteria), bacteria—the Bacteroidetes bacteria (Bacteroidetes), thick wall (Firmicutes), clostridium (Clostridia), Actinomycetes (Actinomycetes), helix (Spirochetes) and warts (Verrucomicrobia) microbacteria flora, but different types of insects there is biggish difference in the composition of intestinal flora.

For the cockroaches, it is divided into omnivorous cockroaches and wood-eating cockroaches. Owing to the different feeding habits of cockroaches, the species of intestinal microbes are different. *B. germanica* and *P. americana* are common omnivorous cockroaches. Studies found that the gut microbiota of *B. germanica* is dynamic due to the omnivorous diets. However, wild and laboratory cockroaches still share common strains in their gut, which constituting the bacterial core of *B. germanica*, and the most abundant families were Porphyromonadaceae, Ruminococcaceae, Rikenellaceae, Lachnospiraceae, Desulfovibrionaceae and Bacteroidaceae (Akbar et al., 2018). Solexa high-throughput sequencing (HTS) technology was conducted on 16S RNA sequences. A total of 23 different phyla and 212 genera were detected. Firmicutes, proteobacteria, bacteroidetes, fusobacteria and planctomycetes are common to the foregut, midgut and hindgut. Acetobacteraceae (proteobacteria) and *Lactobacillus* (Firmicutes) were dominant in the bacterial communities of the foregut and midgut. The abundance of *Enterococcus* in the midgut was higher than in the foregut. The abundance of *Bacteroidete* and *Pseudomonas* was the highest in the hindgut (Zhang et al., 2018). The composition and stability of the different intestinal flora are also different. For the *P. americana*, the hindgut flora is the most stable in different populations, and the intestinal flora changes were not significant after eating different foods (Tinker and Ottesen, 2016). Bacteroidetes, firmicutes and proteobacteria were the predominant phyla present in the gut microbiota of *P. americana* receiving all treatments (Khachatryan et al., 2008). Various protozoan species have also been identified in the gut of German cockroach, which is considered to produces allergens, and four species are typical: *Nyctoterus* sp., *Gregarina* sp., *Amoeba* sp. and *Lophomonas blattarum* (Martã-Nez-Girã et al., 2017).

The hindgut bacteria of wood-feeding termites and cockroaches obviously different from omnivorous cockroaches, and proteobacteria, especially deltaproteobacteria, bacteroidetes, firmicutes, actinomycetes, spirochetes, verrucomicrobia and actinobacteria were predominant, as detected by 16S rRNA (Berlanga, 2015).

3. TRANSMISSION MECHANISM OF SYMBIOTIC IN COCKROACHES

Symbiont transmission is essential to ensure the maintenance of the symbiotic systems through host generations. Two fundamental transmission modes can be distinguished: horizontal and vertical, notwithstanding mixed transmission (Bright and Bulgheresi, 2010). The maternal inheritance and a close physiological relationship with the host make them a useful tool for studies of coevolution (Clark et al., 2001; Yang et al., 2017).

3.1 Vertical Transmission

Insects have structures that facilitate the spread of symbionts, such as crypts. Hosts extracellularly accommodate their symbiotic bacteria either in the lumen of the swollen part of the midgut or in the lumen of the separated sac-like tissues of the posterior midgut called gastric cecum. The bacterial symbiont of *Plautia splendens* is localized within the crypt cavity of the midgut fourth section extracellularly. The symbiont was smeared on the egg surface upon oviposition by adult females, orally acquired by newborn nymphs and thereby transmitted vertically to the next generation and important for growth and survival of the host insects (Hayashi et al., 2015). In urostylidid stinkbugs, the basal region of the ovarioles is enlarged and distended for jelly production, and novel paired organs for symbiont transmission are formed in association with the female genital chamber (Kaiwa et al., 2014).

Vertical transmission usually requires symbiotic to enter the oocytes. Symbiotic bacteria can enter oocytes with haemolymph to ovaries through bacteria-containing cells, be wrapped in bacteria and then enter the ovary. It also can adhere to the surface of the egg and pass to the offspring. Symbionts in Brown planthopper use motor fibrous protein-mediated ovarian pass way, along with blood lymph flows through the epithelial ovarian tube handle sheath when package by it contains bacteria cells, and through cell phagocytosis into oocytes to pass on to offspring (Yukuhiro et al., 2014). In the process of the *Bemisia tabaci* from the nymph to the adult, the cell and molecular remoulding of the bacteria-containing cells have occurred; reduced adhesion between cells, increased cell proliferation, smaller cells to stable sizes and the ability of cells to move autonomously, these changes prompted bacteria-containing cells to carry symbiotic within cells into the ovaries of insects, the vertical transmission of symbiotic was realized (Luan et al., 2016). The gut symbionts of the *Pentatomoidea* are typically transmitted vertically by specific, postnatal transmission mechanisms, such as the bacterial contamination of the egg surface during egg deposition, the excretion of a bacteria-containing mucus or jelly onto the egg mass, and the deposition of bacteria-containing capsules together with the eggs (Kuechler et al., 2016).

There have been some studies on the vertical transmission mechanism of cockroaches' symbionts. Based on a study, a direct mode of microbial transfer from female guts to developing embryos seems unlikely; possible sources of non-Blattabacterium bacteria in embryos besides the gut include the brood sac. Alternatively, the brood sac secretion may be the source of bacterial inoculation from female to developing embryos. Furthermore, embryos may acquire non-Blattabacterium bacteria from the environment in the interval between deposition into the vestibule in the female abdomen and retraction into the brood sac. It is unclear if the vestibule is axenic, but it most likely is not; thus, the brief period between deposition and retraction may be responsible for the presence of gut-associated bacteria in the embryos (Ayayee et al., 2017). Screening of germ-free hatchlings for unculturable contaminants by 16S rRNA gene amplification yielded nothing but the sequences of a *Blattabacterium* sp., i.e., the maternally transmitted obligate endosymbiont found in the fat body of all cockroaches (Lo et al., 2003). Cockroaches, including *Cryptocercus*, harbour endosymbiotic within specialized cells (mycetocytes or bacteriocytes) of their fat bodies; these symbionts, *B. cuenoti*, infect the developing oocyte before vitellogenesis and are maternally inherited (Sacchi et al., 1996). In cockroaches, *Blattabacterium*, as other insect endosymbionts, is transmitted vertically from mother to offspring (Rosas et al., 2018). During transmission, after the ootheca hatch, bacteriocytes containing *Blattabacterium* migrate from the fat body to the ovarioles in nymphs, where they adhere to the follicle cells surrounding the oocyte. After several days, the endosymbionts are expelled by exocytosis as free cells and migrate up to the intercellular space located between the oocyte and the follicle cells, where they are engulfed via endocytosis by the mature oocyte, once vitellogenesis ends and just before ovulation takes place in the adult (around 7 days after adult ecdysis), recovering the host-derived symbiosomal membrane that surrounds each endosymbiont (Giorgi and Nordin, 1994; Tanaka, 2010).

It is well known that wood-feeding cockroaches and lower termites engage in proctodeal trophallaxis, the direct transfer of hindgut fluid from a donor to the mouth of a recipient, the hindgut microbiota are transferred from parents to offspring via proctodeal (anal) trophallaxis (the hatchlings feed on parental hindgut fluids), in order to survive on recalcitrant lignocellulose (Nalepa et al., 2001; Ohkuma et al., 2009). A study aimed at wood-eating cockroaches ultrastructural data suggest that the symbiosis is obligate, and the symbionts are vertically transmitted during *Barbulanympha* cell division and not acquired from the hindgut environment, although the possibility of a free-living stage cannot be completely rejected (Tai et al., 2016). *Barbulanympha* itself is vertically transmitted from adult insects to nymphs together with other hindgut symbionts through proctodeal trophallaxis (the consumption of hindgut fluids directly from the rectal pouch of the donor) (Nalepa, 2015; Nalepa et al., 2001).

3.2 Horizontal Transmission

Horizontal transmission is mainly through mating or feeding on the same kind of food to get bacteria. Faecal transmission and surface contact are also more common. AS1, a new strain of *serratia*, which was isolated from *Anopheles japonicus* with both vertical and horizontal transmission ability, can not only be transmitted vertically to offspring by adhering to the surface of egg shell and laying eggs of female mosquitoes but also be transferred from male mosquitoes to female mosquitoes through mating level (Wang and Qu, 2017). Studies have shown that *Bemisia tabaci* secondary symbiotic, *Wolbachia* and aphids' secondary symbiotic can be transfected between different individuals through the spawning tube of parasitoid wasps (Ahmed, 2015; Gehrer and Vorburger, 2012). The symbiotic Rickettsia can be transmitted from host to parasitic wasp at different nutritional levels through horizontal propagation (Lu, 2016).

As mentioned earlier, cockroaches' neonatal digestive tract is microbe-free and the microbiota becomes fully established by the third instar, at which point the nymphs acquire nutritional independence (Berlanga, 2015; Carrasco, 2014). This is not only the role of vertical transmission but also the role of horizontal transmission, and horizontal propagation will play a greater role in the later stage from nymphs to adults. As *B. germanica* is a nonsocial insect, the lack of proctodeal feeding leaves the offspring with only two possible gut microbiota acquisition: a deterministic transfer by contact with colony members or their faeces, or a stochastic inoculation via environmental factors (Rosas et al., 2018).

Diet is key in hindgut colonization (Engel and Moran 2013; Yun et al., 2014). Previous works indicated that coprophagy is an important behaviour in *B. germanica*, and it has been demonstrated that nymphs consume conspecific faeces (Kopanic et al., 2001). In addition, newly ecdysed first instars nymphs fed only on conspecific faeces survived longer than those deprived of faeces, concluding that adult faeces served as an important nutritional buffer for first instar nymphs (Kopanic et al., 2001). Recently, it has also been demonstrated that aggregation of *B. germanica* is regulated by faecal agents produced (at least partly) by the gut microbial community present in the faeces (Wadakatsumata et al., 2015). By rifampicin treatment of *B. germanica* evidences a faecal transmission route of their gut microbiota, investigator observed that while food and faeces contribute to the gut microbiota acquisition in the cockroach, faeces are the main contributor (Rosas et al., 2018).

Except diet and faecal transmission, some airborne pathogenic microorganisms can be transmitted through the air long distances, due to the popularity of infectious diseases, for a long time in this highly concentrated microbial aerosol environment, the increased infection pressure, can lead to a high risk of infection. We should not overemphasize contact transmission and underestimate the transmission of microbial aerosols, which are mainly transmitted through respiratory tract (Duan, 2008). For example, *Klebsiella pneumoniae* can be transmitted through the air causing extremely high morbidity and mortality (Chandrashekar et al., 1997; Prazmo et al., 2003). And it has been confirmed that *Pseudomonas aeruginosa* can also be transmitted through the air (Jones et al., 2003). Therefore, it can be inferred that the spread of microbial aerosols is also a greater way to spread the symbiosis bacteria.

3.3 Colonization of Intestinal Bacteria During Moulting

The gut microbiota of insects have a unique challenge not faced by those in vertebrate taxa, they experience the process of moulting during the process of growth. Because the insect hindgut lining is ectoderm, it is shed along with the rest of the exoskeleton during the moulting period throughout host development; consequently, insect gut microbiota may require measures to cope with the conditions of the digestive tract during this time. So how do symbiotic colonize cockroaches during their moulting cycle? Does it come from the midgut, which has never shed its skin, or is it derived from the outside?

Taking *Cryptocercus* as an example to explain the desquamate process. Subsocial wood-feeding cockroaches in the genus *Cryptocercus* retain their symbiotic gut flagellates during the host moulting cycle. As *Cryptocercus* enters its moulting period, rising titres of the moulting hormone ecdysone trigger the sexual/encystment (SE) cycles of the flagellates in the gut. During the SE cycle, the protistans become more resistant by producing cysts, pseudocysts or by employing other mechanisms. The lining of the cockroach hindgut is then shed whole, but retained within the host body. The resistant flagellates are then held within the old gut lining (intima) until formation of the new lining is completed. The old intima ruptures, and the protists escape into and repopulate the gut lumen. The old hindgut lining subsequently breaks into small pieces and is voided in faecal pellets along with some resistant protists. Although these protists may pass with host faeces shortly before or after moulting (in some hosts half or more, in others just a few), they play no role in intergenerational transmission under natural conditions. Cleveland described instances of flagellates that do not undergo an SE cycle but are nonetheless retained by *Cryptocercus* during moult. Although in most flagellates all individuals undergo an SE cycle when their host moults, a small percentage of those in the genera *Trichonympha*, *Leptospironympha* and possibly others do not. These individuals are actively motile, free of wood particles and location specific: they are always found in the very anterior part of the hindgut. The number of these individuals that do not undergo, and SE cycle is always sufficient to repopulate the gut after moult (Nalepa, 2017).

Cockroaches are hemimetabolous insects, which pass through several nymphal stages from egg to adult (six in females of *B. germanica*). Gut bacterial load increased from stage n1 to stage n2 but then remained relatively constant through the following moultings, until the adult stage. As early as stage n2, 86% of the total taxa were already present, concomitant with the burst in bacterial abundance per gut. According to the rarefaction curves for each stage of cockroach development, including the ootheca. While both the ootheca and the n1 samples were expected to have the lowest number of taxa, the number in instar n1 was higher than that in the ootheca samples. This indicated that the first colonizers of the gut ecosystem appeared at this stage, even though the relative abundance of *Blattabacterium* was still high. The remaining nymphs and adults had similar numbers of expected taxa. According to Wilcoxon signed-rank tests comparing the indices between nymphs (n2, n3, n4 and n5) and adults, the differences were not significant. Bacterial composition underwent important temporal changes during development, with some taxa showing significant changes in abundance. *Blattabacterium* was the only bacteria detected in embryos inside the ootheca, consistent with a lack of vertical transmission of the gut microbiota. Instead, horizontal transmission of behaviours such as trophallaxis, coprophagy or body/ootheca licking seems probable. Strikingly, the bacterial assembly of n3 samples differed not only from those of adults but also from those of all other nymphal stages. These differences could be related to the physiological state of the host at that particular developmental stage, corresponding to the social/behavioural transitions of the third instar stage. Overall, the gut content among most nymphs is similar but it differs from that of adults. This similarity suggests that at these stages the gut habitat is suitable for a specific fraction of early colonizing microorganisms (Carrasco, 2014).

3.4 Factors Affecting Colonization of Intestinal Bacteria

In terms of intestinal bacteria colonization, host-dominated factors, microbial factors, interactions between microbes and diet may have an effect on colonization of intestinal bacteria (Table 3).

Intestinal epithelial cells are usually covered with mucus layers (main ingredients is mucin) that protect intestinal epithelial cells from microbes on the one hand and provide growth energy for microbes that use them as nutrients on the other (Zhao and Li, 2017; Shan et al., 2018; Zhang et al., 2017). The host-associated microbial communities of the *Panchlora* foregut, midgut and hindgut are each distinct, suggesting that both host

Table 3 Factors Affecting the Composition and Distribution of Bacteria in the Gastrointestinal

Host-Dominated Factors	Microbial Factors	Interactions Between Microbes	Diet
pH, immunoglobulin, salt and enzyme secretion motility such as food emptying rate, intestinal peristalsis, physiology such as septum, scale-like exfoliated cells, mucosin, tissue exudates	Adhesion nutritional requirements spores, faeces, enzymes, antibacterial, substances, reproductive speed	Synergism, metabolic cooperation, growth factor and vitamin secretion, restriction of change in EH, pH, O_2, antagonism or irritation, antagonism or irritation, short chain fatty acids, various amines, antimicrobial substances, siderophore, nutritional competition	Food composition, nondigestible fibre, drugs

association, and the internal environment specific to each gut section, influences community composition (Gontang et al., 2017; Rombout et al., 2014). It can be inferred from above that because of the difference of intestinal environment and the specific symbiotic have special adaptive environment, the specific symbiotic colonize in the specific location of the intestine and its population density is limited by the environment and has a certain range. Wood-feeding *Cryptocercus* and *Parasphaeria* were phylogenetically distant cockroaches, but several bacterial groups were present in both cockroaches and were shared also with termites. Those bacteria may derive from the microbiota of a common ancestor before the diversification of cockroaches, subsequently diversified and adapted in each host. The microbiota depends on the hosts' feeding behaviour and secretions. The composition and physical form of the food changes as it *Cryptocercus* and *Parasphaeria* gut microbiota passes down the gastrointestinal tract, offering microbes at different locations a changing complement of nutrients. Finally, the host obtains multiple nutrients in appropriate quantities and balance to optimally perform their biological function (Berlanga, 2016).

A study of the *Shelfordella lateralis* showed that no observable differences among the gut microbiota of cockroaches fed a low or high-fibre diet (Schauer et al., 2012). Similar works in *P. americana* and the related cockroach species *B. germanica* identified significant changes in their gut communities in response to diet (Bertino-Grimaldi et al., 2013; Pérezcobas et al., 2015; Schauer et al., 2014). Dietary shifts did not result in large changes in gut microbial community composition. No large differences in the relative abundances of major bacterial phyla or families were observed among dietary treatments. *P. americana*'s stable extensive core microbial community appears to be a unique characteristic of the cockroach and is highly resilient to changes in host diet. Insects have evolved diverse mechanisms for the maintenance of stable host–symbiont relationships with their gut microbiota (Khachatryan et al., 2008).

Tegtmeier established the cockroach Shelfordella lateralis as a gnotobiotic model and inoculated germ free nymphs with two bacterial strains isolated from the guts of conventional cockroaches (Tegtmeier et al., 2015). Fluorescence microscopy revealed that both strains specifically colonized the germ free hindgut. In diassociated cockroaches, the facultatively anaerobic strain EbSL (a new species of Enterobacteriaceae) always outnumbered the obligately anaerobic strain FuSL (a close relative of Fusobacterium varium), irrespective of the sequence of inoculation. The exclusive colonization of the germ free cockroach hindgut by strains

EbSL and FuSL suggests that only this gut compartment provides a favourable environment for microbial colonization. This is in agreement with observations of conventional cockroaches, where the hindgut shows the highest density (Tegtmeier et al., 2015). A study uses germ free cockroaches to clearly demonstrate that the assembly of complex gut microbiota in insects can be both predictable and deterministic. Analyses of the normal gut microbiota of cockroaches suggests that the cockroach gut habitat preferentially selects specific bacterial lineages from the environment, that if a preferred lineage is absent from the inoculum, the gut habitat selects other lineages with similar fundamental niches. The gut microbiota of cockroaches is most likely assembled through environmental inoculation, but this process is not entirely stochastic and that the cockroach gut is far from being an environment that indiscriminately accepts bacterial lineages from the environment rather, the gut environment preferentially selects lineages that are specifically adapted to this habitat (Mikaelyan et al., 2015).

4. HOW TO DISTINGUISH BETWEEN BENEFICIAL SYMBIONTS AND PATHOGENS

Insects that are the most widely distributed species in the world live in the environments suffused with a variety of pathogenic microbes. Meanwhile they also harbour large numbers of autochthonous microorganisms peacefully, which have an important effect on life of their host insects via diverse ways and have been important part of insect individuals (Backhed et al., 2005; Stouthamer et al., 1999). So on the one hand, it is essential for insects to resist various exogenous pathogens such as bacteria, fungi, viruses and so on through mounting innate immunity by recognizing specific pathogen-associated molecular patterns (PAMPs). On the other hand, it's also necessary to protect these beneficial resident symbionts from the damage caused by the continuous activation of the immune system through being distinguished from pathogens or weakening the immune reaction, which makes symbionts coexist with hosts and contributes to maintaining homeostasis. Therefore, how to differentiate between beneficial symbionts and pathogens is increasingly becoming our concerned focus problem. Nowadays, most studies on immunology presently focus on how insects defend against exogenous pathogens, but the mechanism of how insects identify and protect beneficial symbionts still remains limited.

4.1 Defence Mechanisms of Insects Against Pathogens

Physical barriers and the activation of innate immune system are the main ways of insects in defence against pathogens. The first defence line of insects against pathogens is the tough cuticle that covers entire body surface. However, when pathogens break through the physical barriers, insects would activate and initiate their immune systems to defend against the invaders, thereby protecting themselves from pathogen infection.

So far, it is believed that in insects there is not acquired immune system similar to that of advanced vertebrates, but nevertheless insects can successfully resist various pathogens in environments thereby succeeding in propagating offsprings, which largely benefits from the well-developed innate immune system. Although the natural immune system does not show a high specificity like the adapted immune system, it has certain specificity because of its capacity of distinguishing between self and nonself as well as between different types of pathogens. At present, it's generally believed that the insect innate immune system can be divided into two arms:

(i) cellular system mediated by various haemocytes, involving defence mechanisms such as phagocytosis, nodulation and encapsulation (Jiravanichpaisal et al., 2006; Li et al., 2013, 2017).

(ii) humoral system resulting in the production of soluble antimicrobial compounds such as antibacterial peptides (AMPs) via intracellular signalling pathways as well as melanin formation and clotting modulated by serine proteases. As a matter of fact, there is no definite boundary between the two arms because they are interrelated and work together to complete multiple immune processes for protecting against the invasion of pathogens within insect body (Feldhaar and Gross, 2008; Zhang et al., 2012, 2015).

Whether insect's humoral or cellular immunity, their first step of initiation both depends on the recognition of insects to potentially harmful microorganisms via the combination between PAMPs and pattern-recognition receptors (PRRs) (Hughes, 2012; Royet and Dziarski, 2007). Invading pathogens could be accurately recognized by PRRs of hosts and rapidly activate insect immune system, which absolutely benefits from the specific structure and chemical nature of its PAMPs. Lysine (Lys)-type peptidoglycan from Gram-positive bacteria, diaminopimelic acid (DAP)-type peptidoglycan from Gram-negative bacteria and β-1–3 glucan from fungi are the major microbe-derived targets for diverse host receptors. PRRs, a kind of host protein responsible for the detection of pathogens and the activation of singalling pathway, could be classified into two main families: peptidoglycan-recognition proteins (PGRPs) and Gram-negative binding proteins (GNBPs) in most insects (Hughes, 2012; Jiang et al., 2010). PGRPs can be a large integral membrane protein with extracellular sensing domain and an intracellular signalling domain; some also can be extracellular soluble proteins or intracellular located PRR proteins. They primarily recognize and bind bacterial peptidoglycans, thereby activating the Toll pathway or Imd pathway to produce AMPs, some of which sometimes may cooperate with each other to enhance the efficiency of pathogen recognition (Chaput and Boneca, 2007; Royet and Dziarski, 2007). GNBPs, the second family of PRRs, bind bacterial ligands such as lipopolysaccharides, lipoteichoic acids or β-1,3-glucans derived from fungi then to initiate intracellular signalling pathway.

Once the invasion signals are identified by the insect's PRRs, subsequently a series of cascade activation reactions involving serine proteases and serpins are elicited and regulated to produce AMPs, melanin and other immune effector molecules, whereby finally eliminating or killing allothigenes. AMPs is a kind of low molecular mass peptides with a net cationic charge, which exhibit a broad range of antibacterial activities and are produced via different mechanisms aiming at different types of pathogens. In insects, there are two major signal transduction pathways to modulate the production of AMPs: one is the Toll pathway evoked by fungi and most Gram-positive bacteria and another is the Imd pathway (immune deficiency pathway) evoked by Gram-negative bacteria. These pathways involve in immune responses of insects through activating different transcription factors to regulate expression of the genes of diverse AMPs. But the two pathways could also cooperate with each other in some responses to resist against certain pathogens together. Besides, Janus kinase/signal transducer and activator of transcription (JAK/STAT) is also a vital signalling pathway in insects to combat pathogens (Myllymäki and Rämet, 2014).

Cockroaches generally live in moist, sunless and unsanitary environments and prefer to feed on fermenting foods, which makes them more easily be exposed to pathogenic microorganisms (Bell and Adiyodi, 1981). As mentioned earlier, insects exclusively rely on the innate immune system to defend against invading microbes mainly via three signalling pathways: the Imd, Toll and JAK/STAT pathways. Recently, a latest study showed that all key components in the Imd, Toll and JAK/STAT pathways as well as effectors, also well express in the *P. americana* genome and in comparison to other insects, many genes in innate immunity, particularly in the Toll pathway, have been extensively expanded. For instance, 12 GNBP1-like and 2 GNBP3-like genes examined in *P. americana* are more than those in any of those insect species. And *P. americana* genome is able to encode 14 Toll proteins while the *Drosophila* genome encodes only 9 Toll proteins. Other components of the Toll pathway, such as easter, spaëtzle, pellino, pelle and cactus, have also experienced gene duplications in *P. americana*. In addition, cockroach AMPs are potentially broad spectrum aiming at various pathogens (e.g., Gram-negative bacteria, Gram-positive bacteria and fungi) and 11 AMPs including defensins, termicins, attacin, drosomycin, prorich peptide (Paprp-1) and antifungus peptide (AFP) have been found in *P. americana* genome. Through RNA interference (RNAi) knockdown of several important genes like Dorsal in the Toll pathway of *P. americana*, researchers concluded that the Toll pathway and AMPs play crucial roles in the American cockroach in fighting invading pathogens. Importantly, Dorsal and other key components in the Toll pathway could be potential molecular targets for controlling this cockroach (Li et al., 2018).

4.2 Immune Reactions to Beneficial Symbionts

Numerous beneficial resident symbionts live with insects in peace, which affect host life in various ways and even are essential part for the host. However, very little is known about the molecular basis of these "peaceful" interactions with the host. Two possibilities are discussed to account for the fact that these symbionts can coexist with the host without eliciting major immune responses: either the symbionts are not recognized anymore as nonself or the symbionts are actively capable to evade the immune system of the host. Interestingly, the genome sequence analyses of several endosymbionts indicate that they have lost many enzymes used for the biosynthesis of regular peptidoglycan and lipopolysaccharide, thereby reducing the visibility of the symbionts by the host immune system (Zientz et al., 2004). Moreover, their obligate intracellular location may further protect them from immune recognition and defence mechanisms.

The digestive tract of the insect may be the most vulnerable part which is not only colonized by many commensal microorganisms but also frequently contacts with a large number of pathogens derived from environments. Thus, there are existing multiple levels of defence systems in the insect's digestive system, which is vital for the moulding of insect intestinal flora and enables insects to identify beneficial symbionts while fighting pathogens.

Taking Drosophila as an example, the intestinal tract of insects is mainly mediated by regulating the expression level of key components in the Imd pathway and the DUOX system, thus obtaining immune tolerance to the beneficial intestinal microflora. In the Imd pathway, the expression of the antimicrobial peptide genes in the intestinal tract is inhibited by the transcriptional inhibitory factor Caudal (Ryu et al., 2008). At the same time, the amidase secreted by the midgut epithelial cells are able to decompose the peptidoglycan (PGN) that causes inflammation, thereby making the PGN secreted by the symbionts maintain at a low level (Zaidman-Rémy et al., 2006). Additionally, the specific PGN-binding receptor in the cytoplasm can be isolated by the Pirk protein, thus reducing the receptors located on the cell surface and delayed the signal transduction of the Imd pathway (Kleino et al., 2008). Another PGN-binding protein PGRP-LE may also play an important role in the immunoregulation of the Drosophila intestinal tract: on the one hand, PGRP-LE can induce the immune response that requires the transcription factor Relish to defend invading pathogens; on the other hand, the immune tolerance of the immune system to beneficial symbionts can also be ensured by upregulating the expression of amidase and Pirk protein (okk). In the DUOX system of the Drosophila, there is also a similar regulatory mechanism. The PGN of bacteria cannot activate the DUOX pathway of the Drosophila alone and it requires ROS (reactive oxygen species) produced by microorganisms to activate the DUOX pathway. When beneficial symbiotic bacteria exist in the gut, the expression of the DUOX gene is inhibited by MKP3, thereby inhibiting the synthesis of the ROS molecule. But the activity of the DUOX enzyme can be induced by intracellular calcium in the Drosophila; so the content of ROS still remains at the baseline level. In recent years, some studies have revealed that the uracil produced by pathogenic bacteria can activate the DUOX pathway in the gut of the Drosophila (Lee et al., 2013). However, intestinal symbionts may evolve to form a mechanism, in which uracil is not generally synthesized in the intestinal tract of insects. Therefore, these symbionts can be distinguished from other pathogenic bacteria, be stable in the insect's intestines and do not excessively activate the immune response. In conclusion, by modulating the two immune pathways Imd and DUOX, insects can effectively identify the symbionts and defend against the harmful pathogens, thereby making the intestines of theirself in normal physiological state.

5. FUNCTION OF ENDOSYMBIONTS

Degradation of plant litter begins well before ingestion by animals, because microorganisms are themselves using the litter as a food source. This is known as microbial conditioning (Swift et al., 1979) and is accompanied by the detoxification of allelochemicals, physical softening of the detritus and the progressive immobilization of N and P in fungal and bacterial tissues as microbial biovolume accumulates. These microbial communities are closely knit, with cross-feeding interactions which make the products and by-products of enzymatic attack by each functional group accessible to others (Costerton, 1992; Mathieu and Sonea, 1996; Shapiro, 1997). Thus, if end product saturation threatens to dampen degradation of the substrate, cooperative microbes may degrade these products to drive the decomposition process forward. Other examples include: (1) physicochemical interactions, like the survival of anaerobic bacteria on surfaces made possible by the consumption of oxygen by aerobes, (2) nitrogen fixation by one species and subsequent sequestration of fixed nitrogen by others, (3) cellulose decomposing bacteria in which nitrogen and carbon compounds produced by the respective microbes are utilized mutually and (4) the products of vitamin synthesis in some strains being used as source compounds for vitamin synthesis by others (Costerton, 1992; Costerton et al., 1995; Wolin and Miller, 1988).

5.1 Nutrition and Metabolism

To a certain extent, the success of insects in diversity and evolution depends on a variety of their relationships with beneficial symbiotic microbes, especially in upgrading nutrient-poor diets (Engel and Moran, 2013). The symbiotic organisms of insects can participate in the nitrogen cycle and provide essential amino acids, also can synthesize nutrients which are necessary to the growth and development of the host but lacking in natural food, such as B vitamins, sterols, amino acids, etc. When the host insects in poor nutrition conditions or the food is difficult to be digested and absorbed, these functions enable the host to better adapt to the environment. Therefore, the symbionts has the extremely vital significance to maintain normal life activities of the host.

5.1.1 Nitrogen Source of the Host

Under limited conditions, many insects rely on symbiotic bacteria to compensate for the lack of nitrogen. Termites feed on simply unspoilt wood, a typical diet deficient in nitrogen. The nitrogen content of woody food is very low, whereas the amount of nitrogen in termites is similar to that of other animals (Fröhlich et al., 2007). It can be seen that termites may have other sources of nitrogen. It is generally thought that termites have the ability to fix and transform nitrogen from the air. Breznak et al. have shown that the nitrogen fixation ability of termites is closely related to the bacteria in the gut (Yang et al., 2006). If termites are fed antibiotics to remove bacteria in the intestinal tract, termites will lose their nitrogen fixation ability, indicating that the nitrogen fixation activity of termites is mediated by bacteria in their guts. In addition, the symbiotic bacteria in the intestinal tract of termites can transform the nitrogen-containing waste produced by the host into high-value nutrients (Wang and Qu, 2017). The nitrogen-fixing bacteria and their metabolites in termite gut may also directly as nitrogen source of growth of termites (Fröhlich et al., 2007). These functions of gut bacteria enable termites to survive on woody food with extremely low nitrogen content.

It is a very effective way that endosymbiont reclaims nitrogen from the host waste products. Unlike other terrestrial insects, cockroach use uric acid as a form of nitrogen storage, rather than a major metabolic waste. And the contents of uric acid in roach may be related to the concentration of nitrogen in the diet (Sabree et al., 2009). Roach usually feed on plants, but also can be omnivorous, feed on animals, even feed on the nitrogen-rich conspecifics (Oliva et al., 2001). When the diet has high concentrations of nitrogen, excess nitrogen can be stored in the fat body cells in the form of uric acid. But when the diet lacks of nitrogen, there is a decrease in uric acid content in roach. The uric acid stored in the roach fat body is considered to be a nitrogen reservoir that can be mobilized when nitrogen is deficient (Patiño-Navarrete et al., 2014).

Several pioneering studies have linked the endosymbionts of roaches with the nitrogen metabolism and the amino acids metabolism of the roaches. It was observed that *Blattabacterium*, the endosymbiont of roach, play an important role. *B. cuenoti* is a species of unique intracellular symbiosis of roaches and primitive termites, which belongs to the class *Flavobacteria* and only lives in the specialized cells of the belly fat body. This endosymbiotic bacteria are thought to be involved in the process of recovering nitrogen from the uric acid, the bacteriocytes have close ties with roach's uricocytes in fat body (Patiño-Navarrete et al., 2014). Studies have showed that *B. cuenoti* can produce urease, suggests endosymbiont may participate in the roach nitrogen economy (Patiño-Navarrete et al., 2014). At the genetic level, it also has been experimentally confirmed that *B. cuenoti* participates in the nitrogen metabolism network of roaches and converts waste nitrogen into nitrogen source (Patiño-Navarrete et al., 2014). At present, the specific process that *B. cuenoti* participate in nitrogen metabolism is not very clear. Some studies suggest that uric acid can be transformed to urea in the roach fat body metabolic system, then the endosymbiont urease would transform urea into ammonia. Ammonia that formed in the process of nitrogen recovery are partly used by endosymbionts to participate in the synthesis of glutamate (González-Domenech et al., 2012). After using antibiotics to remove *B. cuenoti*, the number of mycetocytes decreased significantly, the uric acid was obviously accumulated, and the production of various amino acids decreased (Sabree et al., 2009). The association with *Blattabacterium* can help roach successfully overcome the lack of nitrogen in their diet. Furthermore, *Blattabacterium* is significant to the evolution of roach, and the dependence of the endosymbiont on the Gln provided by the host enhances the obligate characteristic of the symbiotic relationship (González-Domenech et al., 2012).

5.1.2 Carbon Source of the Host

In addition to providing important nutrients to the host directly, symbiosis can indirectly provide nutrients by secreting digestive enzymes that help the host digest food. Cellulose is a rich source of carbon, but it is found in plant cell walls in the form of crystalline or noncrystalline microfibers. Till now, no insect has been found to be able to complete the digestion of lignocellulosic food merely by cellulase and hemicellulase secreted by itself. Therefore, it is difficult for insects, especially the herbivorous insects, to make use of this carbon source. Include insects that feed on wood, such as roaches, termites and carpenter bees, insects that feed on fallen leaves and dead branches, such as silverfish and crickets, and insects that feed on cereals, such as beetles, guts of these insects usually harbours a large number of microbial groups associated with cellulose degradation (Watanabe and Tokuda, 2010), their biodegradation of the lignocellulose is dependent on the combined action of their own and intestinal symbiotic bacteria (Zhang et al., 2017). The help of symbiotic bacteria allows these insects to survive in unbalanced nutritional conditions. Researches have shown that wood-eating roaches and lower termites cannot digest the plant tissue-containing lignocellulose without the help of intestinal symbiotic flagellates. Kill the flagellates in the intestinal tract, the termites died after raising several weeks (Zhao et al., 2012), which confirmed the key role of flagellates in digesting cellulose. The role of different flagellates is not same. Some flagellates mainly degrade highly polymerized cellulose, while some of them digest low molecular weight cellulose (Yang et al., 2006). Flagellates can also hydrolyse cellulose and ferment it into acetic acid, carbon dioxide and hydrogen. Except flagellates, the bacteria that colonized in the hind gut of the termites can produce a variety of glucan enzymes, which help the host decompose cellulose into glucose (Wang and Qu, 2017). In addition, previous studies have isolated some of the bacteria that can hydrolyse hemicellulose from the intestinal tract of the lower termites, suggesting that some bacteria are directly involved in the digestion of the hemicellulose (Yang et al., 2006).

The shift from an omnivorous to a wood-based diet in the ancestor of termites and the *Cryptocercus* lineage was accompanied by the acquisition of cellulolytic flagellates, a specialized gut microbiota and the development of proctodeal trophallaxis (Brune and Dietrich, 2015; Nalepa, 2015). The function of microbes in *Cryptocercus* and wood-feeding termites is to metabolize and upgrade their cellulose-based nitrogen-poor diet (Rosas et al., 2018). In Cryptocercidae, wood is efficiently digested with the help of microbial symbionts housed degradation is attributed either to flagellate symbionts (in phylogenetically lower termites and Cryptocercidae) or to a diverse assemblage of cellulolytic prokaryotes (in the flagellate-free higher termites) in their enlarged hindgut compartments (Brune, 2014; Brune and Ohkuma, 2011). *Bacteroidetes* and *Firmicutes* are also the most predominant phyla in other omnivorous animals, including mammals, revealing their importance in the adaptation to a complex diet, whereas *Proteobacteria* and *Fusobacteria* are shared with wood-feeding termites and *Cryptocercus*, which could indicate the importance of phylogenetically shared common bacteria (Rosas et al., 2018).

5.1.3 Other Nutritional Effects

Intestinal flora also have potential food detoxification function, because many sources of nutrition is only effective under the condition of nontoxic, hydrolysis and some molecules as some plant cell walls into both a detoxification, also can make the food as a useful source of nutrients. A lot of insects specialized to feed on certain poisonous plants, their gut bacteria may poison the sources of food in digestive settlement has played an important role on the (Hehemann et al., 2010).

Insects provide a stable living environment for microorganisms, at the same time, microorganisms participate in nutrition supply and material metabolism of the insects, and effects on growth and reproduction of insects. The symbiotic microorganism plays an important role in changing the feeding habits of insects, balancing the nutritional conditions of insects, expanding the range of insect activities and enhancing the resistance of insects. Using antibiotics to remove the intestinal microorganism of *Blattella germanica* greatly affected the fitness of roach; the relative fitness of roach to remove symbiotic bacteria was only 0.61, which had obvious disadvantage of fitness. In growth and development, the growth period of larva was prolonged and the life span of adult was shortened; in terms of fecundity, the percentage of egg pod shedding increased and the emergence rate decreased, etc. (Li et al., 2016). Plant pests such as aphids, wood pests such as termites, roaches and vector insects of many infectious diseases all have symbiotic microbes. Moreover, many pests have a strong dependence on one or several specialized symbiotic microorganisms. For example, Blattabacterium has important influence on the growth of roaches, termites that lose their body's symbiotic microbes that help digest cellulose will die. The earlier two points provide a new way for the biological control of pests: reducing their living force, breeding force and the capability of carrying the pathogen by reasonably destroying the intestinal flora of the pests, thereby achieving the purpose of controlling the harm of the pests.

5.2 Reproductive Regulation

In the reproductive tissues of insects, there are also many kinds of symbiotic microorganisms, and in general they can be vertically transmitted through the eggs. These symbionts like *Wolbachia*, *Spiroplasma* and *Cardinium* may regulate reproduction of their hosts via various mechanisms, leading to remarkable reproductive abnormalities including cytoplasmic incompatibility (CI), parthenogenesis, feminization, male killing, etc. By these selfish ways, they can significantly enhance efficiency of their transmission among hosts but have negative effects on reproduction of hosts (Werren, 1997; Werren et al., 2008).

Members of the genus *Spiroplasma*, belonging to the class Mollicutes, are wall-less bacteria associated with diverse arthropods (Whitcomb, 1981). In fruit flies, ladybird beetles and butterflies, infection with *Spiroplasma* symbionts often can cause male killing phenotypes, i.e., *Spiroplasma*-infected females produce all-female or female-biased offspring due to male-specific mortality during embryogenesis and/or larval development (Hurst et al., 2003; Hurst and Jiggins, 2000). And some studies showed that host's apoptotic pathway has a vital role in the symbiont-induced male killing (Harumoto et al., 2014). *Cardinium*, belonging to a branch of Bacteroidetes, is a recently discovered kind of symbiotic bacteria that can also cause reproductive abnormalities of its hosts, which is functionally similar to *Wolbachia* (Dumn et al., 2008; Zchori-Fein et al., 2001). But because of the relatively late discovery and narrow distribution range in arthropods, the deepness of research on it is far from *Wolbachia* and it has not been found in cockroaches so far.

In these three kinds of symbionts, *Wolbachia* may be the most widely distributed symbionts in nature and the studies on it are presently most deepening. *Wolbachia*, belonging to α-proteobacteria, are first found in the ovaries of *Culex pipiens* and extremely widespread among arthropod species (Stouthamer et al., 1993). A large number of studies have shown that *Wolbachia* may be most abundant in the insect symbiotic microbial populations, and more than 66% insects including more than 10 orders of insecta are naturally infected with *Wolbachia* (Hilgenboecker et al., 2008). *Wolbachia* are presently found in the following species of cockroaches: *Pycnoscelus surinamensis*, *P. americana*, *Blattella* sp., *Blattella nipponica* and *S. longipalpa* (Vaishampayan et al., 2007). In order to detect the infection of *Wolbachia* in *B. germanica*, Jin et al. utilized one pair of universal primers (81F, 691R) of the *wsp* gene, a gene that encoding the surface protein of *Wolbachia* (wsp), to carry out polymerase chain reaction (PCR) amplification, T-A cloning, sequence determination and homology analysis. As a result, they successfully acquired a 317 bp sequence fragment of the *wsp* gene from the total DNA of the adult *B. germanica* through amplification. And sequence homology analysis indicated that the fragment shares up to 95% high identity with the gene of known *Wolbachia* strain in GenBank, suggesting that *Wolbachia* might widely exist in *B. germanica*, which provides a molecular biological basis for *B. germanica* infection with *Wolbachia* (Jin et al., 2008). Besides insects, there are *Wolbachia* symbiosis also existing in other arthropods like some crustaceans, arachnids and centipedes, as well as most filamentous nematodes (Dong and Wang, 2006).

Wolbachia are maternally inherited obligatory intracellular bacteria (Hilgenboecker et al., 2008). They are mainly distributed in the reproductive tissues of arthropods (e.g., ovaries and testes) and can be vertically transmitted via host's egg cells, which is the basic mode of spreading into the host populations (Stouthamer et al., 1993). During the whole adult stage of female hosts, *Wolbachia* steadily live in their germ cells, thereby being efficiently transmitted to hosts' progenies via the cytoplasm of host's egg cells. In addition to vertical transmission, *Wolbachia* also can be horizontally transmitted among distinct individuals and even between different species, which may be one of the important reasons for *Wolbachia* to break through the limitation on host selection. As *Wolbachia* are unable to leave its host cells to live alone, it is generally believed that horizontal transmission only occurs in the close contact such as wound contact between two hosts.

Numerous studies showed that *Wolbachia* are able to modulate the reproduction of hosts, resulting in diverse reproductive abnormalities, of which CI is the most common phenotype mediated by *Wolbachia*: the eggs that are produced through crosses between *Wolbachia*-infected males and uninfected females hatch abnormally. CI can be divided into unidirectional CI and bidirectional CI. Unidirectional CI is that the progenies produced by the mating of *Wolbachia*-infected males and *Wolbachia*-free females develop abnormally, which finally leads to the early death of their embryos. Conversely, when infected males mate with the females that are infected with the same strain of *Wolbachia* as that of mates, their progenies can develop normally but are also infected with *Wolbachia*. As a result, *Wolbachia*'s infection rate of hosts increases with their propagation and the hosts infected with *Wolbachia* will gradually become the dominant group in the natural population. In addition, if males and females carry different types of *Wolbachia*, bidirectional CI will occur, which make their fertilized eggs die during the embryo period. Thus, the reproductive advantage of the infected females provided by CI leads to the wide spread of *Wolbachia* into host population, which is beneficial to *Wolbachia* but harmful to the host (Dong and Wang, 2006). At present, the molecular mechanism of CI caused by *Wolbachia* has not been known yet, but three models (i.e., lock and key model, removal-restitution model and slow-motion model) can explain the *Wolbachia*-mediated CI phenotype in nature to a certain degree. Because of the complexity of the CI phenomenon, there are still existing certain defects in the three models, which needs to be further studied.

At present, many pest control strategies based on the mechanism of *Wolbachia*-mediated CI have great potential in application of pest biological control and blocking of the spread of vector-borne diseases. And some have been successfully implemented, such as population suppression and population substitution (Pan et al., 2014). The purpose of population suppression based on *Wolbachia*-mediated CI is to reduce the number of target pests. Through constantly releasing a sufficient number of *Wolbachia*-infected males which then mate with target female pests that are not infected with *Wolbachia* or carry different types of *Wolbachia*, the sterility phenotype will spread in mosquito group, gradually reducing and even eliminating target pests. Another is population replacement. Because some studies revealed that *Wolbachia* can also confer mosquito resistant to some human pathogens, including dengue virus, yellow fever virus and malaria parasites, which prevents the growth of these pathogens in mosquitos and reduces insect's vectorial capacity. Therefore, based on *Wolbachia*-mediated CI and *Wolbachia*'s antiviral property, the disease-resistant females that are infected with a new type of *Wolbachia* could be released into the target pest population for pest control. After sufficient release, the new type of *Wolbachia* would quickly spread to all the target populations and as a result, the original wild pest populations in the area would be replaced by a new type of antidisease insect populations, thereby cutting off the spread of vector-borne diseases (Mcgraw and O'Neill, 2013).

Nowadays, with the development of researches on *Wolbachia*, the symbiotic associations between insects and *Wolbachia* have been universally used to control mosquito populations that spread a variety of diseases (e.g., malaria, dengue fever, filariasis and yellow fever), such as the major disease vector species *Aedes* and *Culex genera* (Bourtzis et al., 2014; Daily, 2012). As early as 1967, a large number of bidirectional cytoplasmic incompatible male *C. pipiens* were released into the field to mate with females, effectively suppressing the local *C. pipiens* population, which provides us inspirations for the control of mosquito population by utilizing *Wolbachia* (Laven, 1967). However, *Wolbachia* is only found in *C. pipiens* pallens and a few *Aedes* mosquitoes in natural populations. But heterologous associations could be generated by the experimental transfer of microorganisms from one insect species (or population) into a second species (or population) that does not naturally harbour the microorganism, which solves the problem of limitation on host selection of *Wolbachia*. Scientists have successfully introduced *Wolbachia* from *Drosophila* into *Aedes aegypti* by a two-step process: initial adaptation by serial passage in mosquito cell culture and then injection into *A. aegypti* embryos, leading to stable colonization and vertical transmission. Then they released this kind of *Wolbachia*-infected *A. aegypti* into dengue-endemic regions of several countries, which effectively suppresses dengue virus infection and shortens the lifespan of female mosquitoes, leading to a significant reduction in the vector efficacy of *A. aegypti* dengue transmission (Mcmeniman et al., 2009). Based on this method of mosquito control and mechanisms of *Wolbachia*–host interactions, a new potential strategy could be exploited to suppress pest populations that do not naturally bear *Wolbachia* such as some cockroach populations. We speculate that the species, functions and mechanisms of action of *Wolbachia* within cockroaches are similar to those of mosquitoes. But according to the current investigation, *Wolbachia* are not ubiquitous in all cockroach species or all populations of some species. Some cockroaches have *Wolbachia*, but some do not, which also provides plenty of space and a vast prospect for us to study and utilize *Wolbachia* of cockroaches. First of all, we could find a kind of peculiar *Wolbachia* with extremely strong reproductive ability from some kind of cockroach infected with *Wolbachia*. Then the *Wolbachia* extracted from the cockroach are injected into living cells of another target nature *Wolbachia*-free cockroach, and the cell lines of these living cells were subsequently amplified and cultured in vitro. After serial passage (domestication) and until these *Wolbachia* could infect cockroach cells by themselves, we extract and inject *Wolbachia* into egg pod or embryos of target cockroaches through microinjection, which makes *Wolbachia* be vertically transmitted. Then the *Wolbachia*-infected male progenies produced by these experimental cockroaches are released into fields to mate with native *Wolbachia*-free females. Male cockroaches can mate many times with distinct females during their lifetime, while females mate only once or twice and can produce fertilized eggs throughout their life after one mating. So after mating with the *Wolbachia*-infected males that are artificially released from laboratory, the female cockroaches would never mate with other males and their offspring develop abnormally and die in the embryonic period due to *Wolbachia*-mediated CI, thereby achieving the aim of suppressing the number of cockroaches.

5.3 Defence Function

In throughout life cycle, insects are threatened constantly by various natural enemies such as predators, parasites, parasitoid as well as pathogens. Especially for social insects such as termites, the risk of pathogen infection and the spread of diseases among individuals are greatly increased.

In recent years, more studies have revealed that symbionts make important contributions to enhance the resistance of the insect hosts to many adversity stresses including the damage caused by those enemies mentioned earlier (Oliver et al., 2009, 2010; Xu et al., 2009). And a single insect could defend against various enemies through multiple symbionts residing within its body. In general, there are three different ways in which symbionts can make contribution to their host's defence against these enemies:

(i) Microbes can produce bioactive compounds or their precursors contributing to their host's chemical defence. There are two specific modes including:

Symbionts can synthesize certain chemicals such as various toxins to defend against pathogens or parasitoid directly and protect their hosts from predators, which is the major mode.

Symbionts may secret antimicrobial peptides to protect their hosts from pathogenic microorganisms.

(ii) The insect–symbiont interaction can stimulate or activate the immune system of the host to indirectly strengthen host's resistance to defend against its natural enemies, including pathogens, parasites or parasitoids.

(iii) Besides producing antimicrobial substances or toxic compounds in hosts, symbionts can also directly protect their hosts by competing with the horizontal transmission parasites for host resources, thereby competitively excluding them.

5.3.1 Defence Against Predators, Parasitoids and Parasites

An antipredator defence was verified for the association between rove beetles (*Paederus* spp.) and the endosymbiont, *Pseudomonas aeruginosa* (Kellner, 2002). Compared with the pederin-free larvae, the predation rate of the wolf spiders to beetle larvae that hatched from the eggs containing pederin can be reduced (Kellner and Dettner, 1996). The symbiont *Pseudomonas* in the rove beetle *Paederus* body utilizes the enzymes of the trans-AT PKS family to synthesize the predation which is a kind of potent polyketide toxic substance and can effectively protect beetles from predation, thereby reducing the probability of being preyed on (Dedeine et al., 2001; Kellner, 2002). In addition, some symbionts in termites are capable to produce methane gas to prevent ants or other predators from predation (Rasmussen and Khalil, 1983).

One of the earliest known cases of symbiont-mediated defence in insects involves the protection against parasitoid wasps in aphids. According to some studies, the three γ-proteobacteria endosymbionts: *Hamiltonella defensa*, *Regiella insecticola* and *Serratia symbiotic* can enhance aphids' resistance to parasitoid wasps (Oliver et al., 2010; Vorburger et al., 2010). For example, in the aphid *Acyrthosiphon pisum*, *H. defensa* bacteria provide protection against the parasitoid wasp *Aphidius ervi* (Oliver et al., 2003). Because the presence of the bacteriophage APSE (*A. pisum* secondary endosymbiont) in the symbiont carry toxin genes that encode for the production of toxins to kill the wasp larvae, conferring protection for aphids. Moreover, an alternative strategy for protection by this secondary symbiont may be that parasitoid wasps preferentially lay eggs in *H. defensa*-free aphids (Łukasik et al., 2013a, b).

Few cases are reported about symbiont-mediated defence against nematodes. The only one case that the endosymbiont *Spiroplasma* can enhance the reproductive output of the flies parasitized by the nematode *Howardula aoronymphium* in fly populations occurs in *Drosophila neotestacea* (Hamilton et al., 2014; Jaenike et al., 2010). The presence of *Spiroplasma* can reduce the growth of adult female nematodes in the fly host, eventually leading to the damage to the parasite's fertility and to the reduction of virulence against the host (Jaenike et al., 2010). Although the mechanism has not been known clearly, recent studies revealed that the toxins synthesized by the symbiont *Spiroplasma* may inactivate the ribosomes of parasitic nematodes (Hamilton et al., 2014).

5.3.2 Defence Against Entomopathogens

5.3.2.1 Protections Against Viruses and Pathogenic Bacteria

Besides reproductive regulation function, *Wolbachia* also can enhance resistance to RNA viruses in the fruit fly *D. melanogaster*, and other dipteran insects such as *Culex quinquefasciatus* and *A. aegypti* (Glaser and Meola, 2010). Drosophila C virus (DCV), a nonenveloped RNA virus pathogen that is commonly found in wild populations of *D. melanogaster*, shows high pathogenicity and lethality to flies (Christian, 2005). Interestingly, compared with *Wolbachia*-free fruit flies, the mortality of *Wolbachia*-infected fruit flies was significantly decreased after DVC infection; and the *Wolbachia*-free fruit flies were more susceptible to DCV, indicating that *Wolbachia* can assist flies in fighting DVC to some extent (Glaser and Meola, 2010; Hedges et al., 2008; Teixeira et al., 2008). And the *Wolbachia*-mediated defence against pathogens may be induced by a variety of mechanisms concurrently, such as enhancing the host's innate immune and changing host metabolic pathway and so on.

Gut microbes can help insects resist against pathogenic bacteria. Studies showed that the *Bacillus subtilis* isolated from honeybee gut and honey samples can exhibit antimicrobial activity against an important honeybee pathogen, *Paenibacillus larvae* by surfactin synthesis. *P. larvae* is an extremely contagious and dangerous pathogen of honeybees, which will cause the American foulbrood disease and even lead to rapid death of bees. But the surfactin produced by the *Bacillus* strains can affect vegetative cells of *P. larvae*, thereby controlling bee pathogens biologically and reducing economic loss in apiculture (Sabate and Carrillo LAudisio, 2009).

However, some gut microorganisms can also help pathogenic bacteria to infect hosts. For instance, *Bacillus thuringiensis* (Bt), a kind of pathogenic bacteria of insects, is often used as an important microbial insecticide. Bt infection often causes the damage of the intestinal epithelial cells of hosts to help some intestinal symbiotic bacteria, especially *Serratia* and *Clostridium* enter into haemocoel through host intestinal damage site to proliferate, and then weaken host's immune response, thereby accelerating host death. Because the insecticidal crystal protein produced by BT plays a role only in the alkaline environment and the gut internal environment of *B. germanica* is acidic, Bt is not of high application value in *B. germanica* control (Caccia et al., 2016).

5.3.2.2 The Interaction Between Symbionts and Pathogenic Fungi

In the pea aphid, at least four secondary symbionts (*Rickettsia*, *Rickettsiella*, *Regiella* and *Spiroplasma*) are able to increase survival rates of aphids living with the pathogenic fungi *Pandora neoaphidis*. They also can reduce the formation efficiency of fungal spores, which may reduce the spread of pathogen infection among aphids, thereby improving the inclusive fitness of ahpid population (Łukasik et al., 2013a, b; Scarborough et al., 2005).

In addition, actinobacteria, specifically the genus *Streptomyces* which can produce the most common used antibiotics, help some insects such as digger wasps and fungus-growing ants defend against pathogens through protective symbiosis (Kaltenpoth, 2009). The symbiont *Streptomyces albus* is found in the antennae of female digger wasp belonging to *Philantus* and *Trachypus*. Before spawning, its antennae secrete actinobacteria to the germ cells, which were subsequently transferred from larvae to the surface of the cocoon and then synthesize antibiotics to prevent fungal infection. Similar symbiotic actinobacteria are also found in the leaf-cutter ants *Acromyrmex* and *Atta* (Kaltenpoth et al., 2005). But the inhibitory effect of antibacterial substances produced by the ant *Tetramorium* on the growth of parasitic fungi is still controversial, because the reciprocal fungi are also inhibited (Sen et al., 2009).

At present, considerable numbers of studies have revealed that the insect gut microbiota also plays crucial roles in adjusting the interactions between the host and the intestinal pathogenic fungi. On the one hand, some intestinal microorganisms can help their hosts resist against pathogenic fungi. On the contrary, others are able to assist or promote gut infection through interaction with pathogenic fungi.

The gut microorganisms of *B. germanica* play a vital role in defence against external invasion especially entomopathogenic fungi and promoting the function of immune system (Herren and Lemaitre, 2011; Riparbelli et al., 2012). For example, the germination rate of *B. bassiana* conidia is significantly decreased when passed through the gut of *B. germanica*. The antibiotic-treated *B. germanica*, in which the symbiotic bacteria are eliminated, is more susceptible to entomopathogenic fungal infections, suggesting that the gut microbiota can help the host resist pathogens. Additional studies have showed that the faecal extracts from conventional *B. germanica* is resistant to *B. bassiana*, whereas the faecal aqueous extract from germ-free *B. germanica* is not. Some strains which isolated from the gut of *B. germanica* such as *Pseudomonas* and *Bacillus*, can strongly inhibit the growth of *B. bassiana* in vitro (Huang et al., 2013; Zhang et al., 2013). This may also be an important reason why the fungus infection efficiency through intestinal tract is significant lower that through cuticle.

A series of researches showed that some facultative anaerobic or aerobic bacterial strains isolated from the intestinal tract of *B. germanica*, which belong to Bacillaceae and Pseudomonadaceae including BGI-17 (*Bacillus atrophaeus*), BGI-1 (*Bacillus subtilis*), BGI-14 (*Pseudomonas reactans*) had an obvious inhibitory effect on the growth of the entomopathogenic fungus *B. bassiana* in vitro; one of which showed a strong inhibitory effect and its diameter of the inhibition zone exceeded 20 mm. And the inhibitory effect of these strains may result in the sick vacuolar enlargement of fungal hypha, the leakage of protoplasm and then the shrinking of some sick vacuoles, eventually leading to the fracture of hyphae. These experiments indicate that these gut bacteria may play vital role in resisting the invasion and the colonization of external pathogenic fungi in German cockroach (Huang et al., 2011; Huang et al., 2013; Zhang, 2012).

In order to explore the interaction mechanism between entomopathogenic fungi and the intestinal fora of *B. germanica*, the composition of the intestinal flora in cockroaches was analysed through HTS. After oral infection by *M. anisopliae*, the mortality of normal cockroaches and the cockroaches that are removed intestinal flora is significantly different and the mortality of the cockroaches without intestinal flora is notably increased. Then 13 antifungal strains were isolated from the cockroach gut. The results of the bioassays on these 13 bacterial strains isolated from foregut, midgut, hindgut and faeces, respectively, occupied 23.1%, 30.8%, 15.4% and 30.8%, which showed that the gut microbiota, especially those in the hindgut, may make a significant contribution to protect cockroaches from fungal invasion. The strongest antifungal activity in the hindgut samples was suspected that due to the high diversity and greater abundance of gut microbiota, especially bacteroides and pseudomonas. This study explains that an important reason why the efficiency of *M. anisopliae* infection with *B. germanica* via intestinal tract is very extremely low may be related to intestinal bacteria (Zhang et al., 2018).

On the other hand, when pathogen epidermal fungal infection breaks through its immune mechanism, some opportunistic pathogens may assist pathogen fungal infection and accelerate insect death. For example, through observation and experiments, Wei et al. found that after topical fungal infection, the death time of the mosquitoes with gut microbiota is significantly shorter than that of the mosquitoes whose microbiota is removed. Moreover, because of fungal infection, the gut bacterial load increased significantly but the bacterial diversity decreased greatly, thereby leading to the microdysbiosis of mosquito gut microbiota. According to further study, the entomopathogenic fungus *B. bassiana* can interact with the gut microbe *Serratia marcescens* that will proliferate in the midgut and then translocate to the haemocoel due to fungal infection, to promote fungal killing of mosquitoes by downregulating the mosquito midgut immune responses including the expression of antimicrobial peptide and dual oxidase via secretion of the toxin oosporein (Wei et al., 2017).

Interestingly, studies found that the Madagascar hissing cockroach mite, *Gromphadorholaelaps schaeferi*, prevents fungal infection in its cockroach host, i.e., the Madagascar hissing cockroach *G. portentosa* by involving in a cleaning symbiosis thereby prolonging the lifespan of its host. *G. schaeferi* is a permanent symbiont of cockroaches and is obtained by cockroaches from their mothers. The mite *G. schaeferi* can reduce cockroach surface moulds by removing the food debris from the cockroach that is used as a substrate required for the growth of fungi, which improves the survival rate of *G. portentosa*. In addition, limiting the food supplied to the cockroaches that results in the absence of saliva and food debris, moist food residue that comes from cockroach's chew on the cockroach bodies is observed to have a synergy effect with the cleaning symbiosis relationship between mites and cockroaches, thereby further discouraging the growth of most severe allergenic cockroach moulds.

Based on this study, we can suppress the total amount of cockroach mould allergen and restrict the mould growth as well as mould load on the Madagascar hissing cockroaches by keeping *G. schaeferi* cleaning symbiosis on the cockroaches in combination with limiting food input into the cockroach colony, which both lead to less saliva and moist debris accumulated on the cockroach's body. Additionally, if we could reduce the simple sugar load and moisture level of the cockroach's body by using dry food instead of moist, fresh fruits and vegetables for feeding cockroaches, the mould load and the presence of medically significant strains would be significantly suppressed. Therefore, this will help us reduce associated allergenic diseases caused by cockroaches.

5.4 Insecticide Resistance

Nowadays, insecticide resistance caused by insecticide abuse for pest control is increasingly universal, which has globally drawn people's attention. So people began to explore the mechanism of insecticide resistance and most studies focus on physiology and biochemistry mechanisms (Ruan et al., 2012). Nevertheless, because few researchers carried out pesticide resistance experiments with axenic insects, they all ignored an important issue: just as the human gut microbiota modifies drug efficacy, insect symbionts can also affect how pesticides work (Sousa et al., 2008).

Symbionts also possess the ability to assist insects in degrading xenobiotics and confer insecticide resistance on most kinds of pests. At present, increasing cases of symbiont-mediated xenobiotic especially insecticide degradation in insects have been found (Gressel, 2018). According to current researches, the mechanisms of symbiont-facilitated pest resistance involve the following two main aspects.

First, pesticides can induce symbionts to encode enzymes for detoxification, which are able to metabolize insecticides into one or more metabolites which then can be absorbed by insects through mineralization or cometabolism, ultimately achieving insecticide detoxification in insects.

Studies have revealed that there are significant difference in activities of the enzymes synthesized by symbionts, such as carboxylesterase (CarE), mixed-function oxidase (MFO) and glutathione-*S*-transferase (GST) in two strains (imidachloprid-resistant strain and susceptible strain, $P < 0.05$) of *Nilaparvata lugens*, which indicates that the symbionts of *N. lugens* can contribute to host resistance to imidachloprid through enhancing activity of CarE and MFO (Li et al., 2010).

In addition, Japanese scientists have demonstrated that the genus *Burkholderia* and allied stinkbug species could establish a beneficial symbiotic relationship with *R. pedestris* and enable the host to resist fenitrothion (Kikuchi et al., 2011, 2012). Under natural conditions, there is a little amount of *Burkholderia*, the fenitrothion-degrading symbiotic bacteria in the farm soil. But on account of extensive repeated applications of fenitrothion to field soils, fenitrothion-degrading bacteria were sharply enriched abundantly (Hayatsu et al., 2000; Singh, 2009; Tago et al., 2006). Then the bean bug *Riptortus pedestris*, a leguminous crops pest, environmentally acquires the Burkholderia symbiont from field soil through horizontal transmission at its nymph stage, which is similar to the acquisition way of *Rhizobium* by leguminous plants. And *R. pedestris* quickly establish a special mutualistic symbiosis with *Burkholderia* in the posterior part of host midgut. The bacteria are capable of hydrolysing the organophosphorus compounds fenitrothion into 3-methyl-4-nitrophenol with only little insecticidal activity and decompose the degradation product as a carbon source for their growth, which may accelerate biodegradation of fenitrothion in the environment and confer insecticide resistance on *Riptortus pedestris* (Kikuchi et al., 2005, 2011a; Schaefer and Panizzi, 2000). Similarly, the gut symbiont *Citrobacter freundii* BD (CF-BD) can confer the pest oriental fruit fly *Bactrocera dorsalis* insecticide resistance to trichlorphon by upregulating phosphatase hydrolase genes of itself, thereby impacting pest management (Cheng et al., 2017). Recently, a latest study shows that *Stenotrophomonas maltophilia* OG2 that is isolated from the intestine of cockroaches living in pesticide-contaminated environment, is able to degrade α-endosulfan by various enzymes via a hydrolysis pathway in nonsulphur medium as a sole sulphur source for growth within 10 days of incubation (Ozdal et al., 2017).

Second, endosymbionts can activate the immune system of the host in order to achieve the detoxification of insecticides. For instance, *Enterococci* can strongly induce the expression of detoxifying enzymes such as GST and CarE to improve host resistance to chlorpyrifos in *Plutella xylostella* (Xia et al., 2013).

However, whether endosymbionts participate in the resistance of *B. germanica* to pesticides cannot be determined at present. In order to explore the relationship between the composition of the intestinal flora of *B. germanica* and its insecticide resistance, we used resistant strains (RR) and susceptible strains (SS) of *B. germanica* to cypermethrin in our laboratory as object of this study and analysed flora of the intestinal flora in the anterior, middle and posterior gut of the two *B. germanica* strains by illumina HTS technology. Then the different strains in the gut of the two strains (RR and SS) were isolated and identified in vitro, and their resistance mechanism was preliminarily explored in three aspects: the degradation of cypermethrin in vitro, the effect of intestinal strains back to the cockroaches on the resistance of test insects and the changes of glutathione-*S*-transferase (GST) and carboxylesterase (CarE) activity. The results showed that there are 23 bacteriophyte and 212 bacterial genera in the gut of *B. germanica*, among which five genera of the two cockroach strains with the highest relative abundance were *Acetobacteraceae*, *Lactobacillus*, *Weissella*, *Alistipes* and *Bacteroides*. In the foregut and midgut of the two strains, the composition of their dominant flora is basically identical, but there was a great difference in abundance. In the resistant strains, the amount of *Acetobacteraceae* and *Lactobacillus* was notably decreased while the abundance of *Weissella* was significantly increased in contrast to susceptible strains, indicating that insecticide treatment may have a selective effect on the gut bacteria of *B. germanica*, such as killing the sensitive bacteria. It is suggested that the changes of intestinal bacterial flora of *B. germanica* may play certain role in pesticide resistance. But in hindgut, the difference whether in dominant flora or in relative abundance was not significant for the two strains. Degradation experiments in vitro as well as resistance monitoring experiments in vivo revealed that intestinal symbionts can markedly enhance the insecticide resistance of insects. Preliminary studies have shown that *Enterobacter* sp. may directly reduce the toxicity of cypermethrin by the way of degradation in vitro, but this is not the only mode of action. Additionally, *Weissella* may facilitate insecticide resistance by stimulating the immune responses of insects to improve the activity of detoxifying enzymes. Furthermore, the results of enzyme activity tests showed that the activity of different detoxifying enzymes can be enhanced by distinct intestinal symbionts. *Enterobacter* sp. and *Weissella* back to gut could enhance the activity of CarE and GST (two kinds of detoxifying enzymes), respectively, further suggesting that the gut microbiota composition of *B. germanica* may be related to insecticide resistance (manuscript in preparation).

Based on the earlier four experimental results, we can conclude that the gut bacteria can cause changes in insecticide resistance, which is mainly carried out through two mechanisms: direct degradation and stimulating the immune responses of insects. As a whole, this study revealed a new mechanism of insecticide resistance and identified active strains that can degrade pesticides from the gut of the cockroach. This not only sheds light on a perspective for comprehensive management of pest insecticide resistance but also likely provides feasible strategies for environmental restoration. For example, if we could destroy the normal growth of intestinal flora in the body of *B. germanica* by preparing the microecologic agent or using antibiotics, the insecticide resistance level of the German cockroach would be dramatically reduced, thereby achieving the purpose of controlling *B. germanica*. Combined with this earlier method, only using small doses of insecticides can eliminate pests effectively and rapidly, which also contributes to environmental protection at the same time.

5.5 Effect on Behaviour

Symbionts within insects are also involved in the modulation of insect behaviours. Symbionts are able to synthesize or decompose compounds in insects to produce metabolites used to synthesize pheromones or interspecific hormones, thereby affecting the behaviour of insects. Also, some symbionts can modify the body colour of insects to affect hosts' relative susceptibility to predators and parasites (Xu et al., 2009).

In general, predators and parasitic insects identify prey mainly based on visual cues, thus body colour is an ecologically identification characteristic associated with predatory insects (Leonard and Cordoba-Aquilar, 2010). In nature, there are red and green pea aphid *Acyrthosiphon pisum* existing simultaneously in the same population. Previous studies showed that the body colour of aphids is genetically controlled and the number of red aphids is much more than that of green aphids (Müller, 1962). Colour variation within natural aphid populations varies with the predation and parasitism pressures (Ruxton et al., 2004). Ladybird beetles tend to easily find red aphids on green plants, while parasitoid wasps prefer to attack green aphids (Libbrecht et al., 2007; Losey et al., 1997). Recent studies have showed that members of the genus *Rickettsiella* can alter the body colour of the pea aphid *A. pisum* from red to green by inducing aphids to synthesize green pigments. For aphids, this change of body colour makes contribution to avoiding predators who are inclined to prey on red ones, such as ladybird beetles. However, the resistance of green aphids to parasitoid wasps is reduced. Interestingly, within the bodies of these aphids, *Rickettsiella* always coexists with *Hamiltonella* or *Serratia*, thereby indirectly protecting them from parasitoid wasps (Oliver et al., 2010).

In recent years, the close relationship between gut microbes and insect behaviour has been increasingly discovered and clarified, which is of great significance for the development of new pest control strategies. For example, *Pantoea agglomerans*, the gut microbial bacteria of *Schistocerca gregaria* can make use of the lignin derivative vanilla acid of the digested waste products from the locusts to synthesize guaiacol. And guaiacol is the precursor of the aggregation pheromone that can promote the locust swarm behaviour through faecal volatilization (Dillon et al., 2000). Gut microbes can not only affect insect behaviour by synthesizing pheromones but also affect mating preference or preferences of insects. The fly gut symbiotic bacteria such as *Lactobacillus plantarum* were responsible for mating preference of *D. melanogaster* by changing the levels of cuticular hydrocarbon sex pheromones. The fruit flies strongly tend to mate with the heterosexual individuals whose gut microbiota composition is similar to themselves, which may eventually lead to the formation of the new species (Sharon et al., 2010).

Recently, a crucial factor of affecting aggregation of *B. germanica* has been found: the gut microbes contribute to the production of the faecal volatile carboxylic acids (VCAs), a kind of chemical pheromones that promote the aggregation behaviour of *B. germanica*. Through chemical analysis of the faecal extract, researchers found that there are 40 kinds of highly attractive VCAs in the faecal extracts of normal *B. germanica* whereas only relatively unattractive 28 kinds in the faeces of axenic cockroaches (no microorganisms in the digestive tract); among them, the amount of 24 kinds was significantly decreased compared with the normal *B. germanica*. Then they revealed that nymphs like the faecal extract of normal cockroaches much more than that of axenic cockroaches through behavioural bioassays. To further study the function of gut bacteria in aggregation of *B. germanica*, they cultured, isolated and identified aerobic bacteria from faeces of *B. germanica* and inoculate them into cockroaches, respectively. As a result, the aggregation activity is recovered in the axenic cockroaches inoculated with faecal aerobic bacteria through the diffusion of highly attractive faecal VCAs. Additionally, when the six bacterial strains (*Enterococcus avium*, *Weissella cibaria*, *Pseudomonas japonica*, *Pseudomonas monteilii*, *Acinetobacter pittii* and *Acinetobacter* sp.) isolated from the faeces of normal *B. germanica* were all inoculated into axenic *B. germanica*, the aggregation response of the cockroaches was stronger and more effective as compared to the cockroaches inoculated with single isolates (Wadakatsumata et al., 2015). This study indicated that some gut microorganisms in cockroaches can contribute to the biological synthesis of pheromones, further facilitating aggregation behaviours.

As is known to all, cockroaches usually live gregariously (Rust et al., 1995). Aggregation could help the individuals in population contact with each other and contributes to the development of nymphs who have coprophagy and the sexual maturity of cockroaches (Kopanic et al., 2001; Uzsak and Schal, 2013; Uzsák and Schal, 2012). Besides, aggregation behaviour plays a vital role in various aspects of cockroach life, such as predator defence, mate location, thermoregulation, prevention of water loss and so on, which significantly increases the fitness of cockroaches in the changing human environments. And their group foraging may also be facilitated by rapid and accurate aggregation behaviour even if their life environment varies frequently. If the aggregation behaviour of cockroaches was artificially destroyed via certain methods, the gregarious cockroaches will transform to the solitary cockroaches, of which the vitality will be greatly decreased, thereby achieve the aim of cockroach control. Based on the vital stimulative role of symbionts in the synthesis of insect pheromones, new potential strategies for cockroach control is urgent to be explored and put into application. For example, we can interfere with the production of pheromones associated with aggregation by disrupting the cockroach gut microbial homeostasis, thereby resulting in dissolution of their aggregation, thereby effectively controlling cockroaches. In addition, some researchers successfully inhibited the aggregation of solitary Locusta *Migratoria manilensis* by making use of the microsporidian parasite *Nosema locustae. N. locustae* can suppress the growth of hindgut microbes that involve in the synthesis of aggregation pheromones by modulating the locust immune system and altering the chemical properties of the locusts' gut such as acidifying the hindgut, thereby further disrupting the nervous system of the locusts. As a result, the productive levels of neurotransmitters (serotonin and dopamine) that initiate and maintain locust gregarious behaviour are significantly decreased, resulting in gregarious locusts back to the solitary phase (Shi et al., 2014). Similarly, we can also utilize the microsporidian parasite *Nosema* to infect cockroaches, which will cause the body consumption and disturb the aggregation behaviour of cockroaches, ultimately reducing the fitness of cockroaches and effectively controlling them.

5.6 Other Functions

Insects are poikilotherm who are very sensitive to temperature changes. The adaptability of insects to high temperature is an important factor of affecting their geographical distribution and spread. When the temperature changes exceed a certain range, various physiological reactions even death will happen to insects (Cui et al., 2008; Du et al., 2007). Most insects respond to temperature stresses by regulating some metabolic pathways or upregulating the expression of resistant substances like heat-shock protein (HSP) (Lü and Wan, 2008; Mahadav et al., 2009; Salvucci et al., 2000).

However, abundant studies have shown that symbionts also play an unnegligible role in enhancing temperature tolerance of insects, thereby making insects evolve in a good direction (Feldhaar, 2011). At present, the mechanism of symbionts' contribution to the temperature tolerance of insects has not been fully clarified. Based on some crucial research achievements, it could be inferred that there are several mechanisms as follow.

First, some symbionts may enhance insect adaptation to high temperature by modulating insect metabolism. For instance, at high temperature, some metabolites such as fructose, mannose-6-phosphate (M6P) and *N*-acetylglucosamine may be increased notably in the pea aphids that carry *Serratia*, which would help insects to resist the adverse effects of high temperature (Burke et al., 2010). In addition, some genes of symbionts may directly express to produce relevant proteins or involve in regulating the expression of substances related to stress resistance in hosts, thereby enhancing heat tolerance of hosts. For example, *Buchnera* is the primary endosymbiont of the pea aphid and some of its genes such as *ibpA* involve in the expression of the proteins related to host insect resistance to heat. When host is in a high temperature environment, the expression quantity of *ibpA* that can encode a kind of HSP is rapidly increased. After mutation of the *ibpA* gene, the mutant aphids cannot produce offspring or produce very few offspring at high temperature (Burke et al., 2010; Dunbar et al., 2007). Besides, at normal temperature, the expression of some microfilament and cytoskeletal genes, such as actin gene, tropomyosin gene, which all have a potential role in the resistance of *B. tabaci* to high temperature, in the *Bemisia tabaci* populations that carry the symbiont *Rickettsia* was relatively high (Brumin et al., 2011). Furthermore, some secondary endosymbionts may indirectly maintain the adaptability of hosts by protecting primary endosymbionts at high temperature. For example, the bacteriocytes of the pea aphid *Acyrthosiphon pisum* are susceptible to temperature, while the presence of secondary symbionts, Pea aphid *Serratia symbiotica* (PASS) and Pea aphid *Rickettsia* (PAR), can reduce the damage caused by high temperature. And *Buchnera*, the primary endosymbiont of aphids that could promote the heat tolerance of insects, exists in these bacteriocytes (Montllor et al., 2002). Afterward, some studies revealed that it's *S. symbiotica* that protects *Buchnera* from high temperature in the host (Burke et al., 2010).

In addition, the presence of some symbionts can also enhance host's adaptability to low temperature. For example, most of the leaf-cutter ants, which generally live in warm and humid tropical regions, are provided with food via symbiosis with the Agaricaceae fungus *Attamyces* who is considerable strict with requirement for temperature and usually grows slowly at below 20°C. However, this symbiotic strain of the leaf-cutter ant *Atta texana* that live in the north of America has strong cold tolerance and the leaf-cutter ant can enhance adaptability to low temperature through maintaining the growth of the fungus *Attamyces* in winter (Mueller et al., 2011).

Therefore, as for the insect species under the environmental selection pressure such as unfavourable temperatures, symbionts may act as a "modulator" to form holobiont with host insects, which has an evolutionary advantage under adverse conditions, thereby improving the adaptability of insects to the adverse environments (Shan et al., 2016).

6. OUTLOOK

Cockroaches are semisocial insects which usually live gregariously and prefer to live in damp and sunless areas where microbes are easy to reproduce. Moreover, they crawl very fast and have coprophagia, which is easy for them to be exposed to and acquire various microorganisms, as well as quickly spread these microorganisms into their populations (Gore and Schal, 2007). So the utilization of symbionts to control cockroaches may have a great potential. In conclusion, symbionts are ubiquitous in cockroaches and have pervasive impacts on multiple characteristics of the host. Cockroach-associated symbionts must be considered as a method to effectively control cockroaches.

6.1 Application of Wolbachia in Cockroach Control

Wolbachia are endosymbiotic α-proteobacteria that are extremely widespread among arthropod species (Dong and Wang, 2006). For cockroaches, Wolbachia are present in *Blattella* sp. and *S. longipalpa* (Vaishampayan et al., 2007). Wolbachia has great potential in insect cockroaches control and prevention of insect-borne diseases due to the function of the reproductive regulation in many insects. At present, *Wolbachia* has been successfully used to control mosquito populations through CI, including population suppression and population substitution (Pan et al., 2014).

Based on this method of mosquito control and mechanisms of *Wolbachia*–host interactions, a new potential strategy could be exploited to suppress pest populations that do not naturally bear *Wolbachia* such as some cockroach populations. We speculate that the species, functions and mechanisms of action of *Wolbachia* within cockroaches are similar to those of mosquitoes. But according to the current investigation, *Wolbachia* are not ubiquitous in all cockroach species or all populations of some species. Some cockroaches have *Wolbachia*, but some do not, which also provides plenty of space and a vast prospect for us to study and utilize *Wolbachia* of cockroaches. First of all, we could find a kind of peculiar *Wolbachia* with extremely strong reproductive ability from some kind of cockroach infected with *Wolbachia*. Then the *Wolbachia* extracted from the cockroach are injected into living cells of another target nature *Wolbachia*-free cockroach, and the cell lines of these living cells were subsequently amplified and cultured in vitro. After serial passage (domestication) and until these *Wolbachia* could infect cockroach cells by themselves, we extract and inject *Wolbachia* into egg pod or embryos of target cockroaches through microinjection, which makes *Wolbachia* be vertically transmitted. Then the *Wolbachia*-infected male progenies produced by these experimental cockroaches are released into fields to mate with native *Wolbachia*-free females. Male cockroaches can mate many times with distinct females during their lifetime, while females mate only once or twice and can produce fertilized eggs throughout their life after one mating. So after mating with the *Wolbachia*-infected males that are artificially released from laboratory, the female cockroaches would never mate with other males and their offspring develop abnormally and die in the embryonic period due to *Wolbachia*-mediated CI, thereby achieving the aim of suppressing the number of cockroaches. In fact, the parthenogenesis function of *Wolbachia* also can be used to prevent pests. *Wolbachia* which has parthenogenesis function not only induced parthenogenesis by affecting the meiosis of host cells but also decreased the female's egg laying (Pannebakker et al., 2004).

6.2 Interaction Between Symbionts and Entomopathogenic Fungi

In host insects, the interaction between symbionts and entomopathogenic fungi can provide new ideas for the prevention and treatment of cockroaches.

Symbionts can produce certain chemicals, such as toxins, that act directly on entomopathogenic fungi and bacteria. For example, female adults of *Philanthus train gulum* carry symbionts *Streptomyces* spp. that can produce a "cocktail" of nine kinds of antibiotics, so as to protect the larvae against lethal fungi and bacteria (Shokal et al., 2016). The same situation occurs in termites and bees (Kroiss et al., 2010).

The gut microbiota of *B. germanica* is closely related to the defence of entomopathogenic fungi. There have been numerous studies showing that gut microbiota can help host resistance to entomopathogenic fungi. For example, germ-free cockroaches are more sensitive to pathogenic fungi such as *B. bassiana* than conventional cockroaches. It is found that due to the high diversity and large number of gut microbiota, especially *Bacteroides* and *Pseudomonas* in the hindgut, may play an important role in protecting cockroaches from fungal invasion (Zhang et al., 2018).

With the rapid spread of serious pesticide resistance, it is possible to control the cockroach with entomopathogenic fungi based on the study of the interaction between symbionts and entomopathogenic fungi. The oral infection rate of fungus insecticide is extremely low due to the gut microbiota, if we are able to use potential synergy agent with stomach poisoning functions such as boric acid and abamectin, as well as cockroach pathogenic microorganism to break or change cockroach's microecological composition, then the practical applications of entomopathogenic fungi would be greatly improved. There have been numerous reports of synergistic effects of *M. anisopliae* and boric acid, *M. anisopliae* and serratia. It is speculated that the synergistic effects mechanism is related to the destruction of gut structure and the compositions of microbiota (Zurek et al., 2000). In fact, the combination of commercial formulations (chlorpyrifos, fipronil and chlorantraniliprole) and *M. anisopliae* to control the adult *P. americana* can achieve a higher mortality rate of cockroaches than those used alone (Wakil et al., 2012). This integrated approach will improve the toxic effect of *M. anisopliae* as a kind of biopesticide for controlling cockroaches and reduce the kill time to cause the similar lethal effect in comparison to entomopathogens alone. This could also prolong the use of specific insecticides by reducing the total amount of pesticides used, reducing the harm to human health and the environment.

6.3 Paratransgenesis

Paratransgenesis, a development of synthetic biology, and the study of insect symbiosis have developed a new pest control strategy. The basic principle of paratransgenesis is through genetic manipulation of the insect symbiont, which can be used as gene expression vector to produce transgenic effect on the host, and to realize the prevention and control of pests and insect-borne diseases. The key requirement for paratransgenesis is those symbiotic microorganisms are cultivable and amenable to genetic manipulation and are easily propagated in insects to facilitate the propagation of desired traits.

The technology has been used as a promising prevention tool to stop the spread of vector-borne diseases. Trypanosomiasis was caused by *Trypanosoma cruzi*, a pathogenic microorganism in triatomine body. The transgenic bacterial symbiont *Rhodococcus rhodnii* expresses a functional antibody fragment that is exported into the gut lumen of *R. prolixus*, an important vector of *T. cruzi*, and etiological agent of chagas disease, which exhibits effective inhibitory activity against *T. cruzi* by expressing an anti-trypanosomal protein in the gut of the triatomine bug (Durvasula et al., 1999). In addition, *R. prolixus*'s population is susceptible to infection when they feed on simulated triatomine faecal pellets that contain genetically modified bacteria in the domestic environment. Analogous approaches can be used to control of the *Leishmania*, which parasite by *Phlebotomus argentipes*. A natural symbiotic bacterium (*Pantoea agglomerans*) in the midgut of *A. mosquitoes* was targeted for genetic modification, and the *Escherichia coli* haemolysin A secretion system secretes a variety of anti-*Plasmodium* effector proteins (SM1, scorpine, etc.) by *P. agglomerans*; it is found that the development of *Plasmodium falciparum* in the mosquito was efficiently suppressed when paratransgenesis was used to deliver antimalaria effector molecules to the mosquito midgut (Abraham, 2017). This strategy provides a new way for the use paratransgenesis as a powerful tool to control the transmission of vector-borne diseases.

Paratransgenesis can be applied to the prevention and control of cockroaches. *B. sphaericus* A3–2 and *B. cereus* is the pathogenic bacterium of cockroach, and it has insecticidal activity to *B. germanica*. Sphaericolysin, a kind of toxic protein produced by *B. sphaericus* A3–2, can make the *B. germanica* paralysed quickly at doses of 100 ng, also can rapid pyrolysis blood cells, causing the death of the insect body (Nishiwaki et al., 2007). Sphingomyelinase C, a kind of toxic protein produced by *B. cereus*, can degrade the phospholipids in the cell membranes of neurons, and lyse the cells, causing the death of the insect body. However, *B. sphaericus* A3–2 and *B. cereus* do not exist in the cockroach. They must break through the immune system to play the role of killing the cockroach, when they enter the body of the cockroach, but most are destroyed. Therefore, paratransgenesis plays an important role in the prevention of cockroaches. *B. subtilis* is the resident bacteria in cockroaches, it can be used as the target of genetic modification. First, separating *B. subtilis* from the gut of cockroaches; second, transferring sphaericolysin or sphingomyelinase C into *B. subtilis* by transgenic; third, returning *B. subtilis* to the cockroach, making it produce this two kinds of toxic protein in the cockroach (Fig. 1B) (Arora and Douglas, 2017). Through transgenic symbiosis, *B. subtilis*, which is not recognized by the immune system, can produce toxin proteins to control cockroaches. Other toxic protein produced by nematodes, fungi and so on can also be transgenic into cockroaches.

6.4 Pesticide Resistance Management and Push–Pull Strategy

Symbionts play a vital role in degrading pesticide and confer insect insecticide resistance. They help insects degrade pesticides in the two major ways, producing detoxifying enzymes and activating the immune system of the host. Based on this, systemic fungicides and antibacterial compounds could act as pesticide synergists to prevent or overcome symbiont-facilitated pesticide resistance.

Abuse of pestcides frequently makes severe resistance, so integrated pest management (IPM) must be adopted for pest control. IPM involves various effective strategies: physical control, chemical control, biological control and so on and achieves the goal of pest control at the lowest economic cost and the minimum environmental damage.

As important part of IPM, the push–pull strategy has increasingly people's attentions because it emphasizes the utilization of natural factors such as natural enemies, which could significantly avoid environment pollution. The basic principle of the push–pull strategy is the integrated utilization of the substances that can influence pest behaviour so as that alter the distribution and density of pests. It includes two main aspects: using repellents to remove pests from crevice where the pesticide is difficult to handle (push); using attractants to attract pests into an area where the pesticide is easy to handle (pull) and then intensively eliminating pests.

Because the distribution of cockroaches is secluded and dispersive, it is more necessary to make full use of the push–pull strategy to remove cockroaches from the areas where pesticide is difficult to handle or other areas where pesticide is easy to handle, so that they can be intensively eliminated by other methods like chemical control. The pheromone of *B. germanica* can be used in the push–pull strategy, mainly including sex pheromone, aggregation pheromone, dispersing pheromone. Ishii and Kuwahara found that aggregation pheromones of *B. germanica* mostly produced by rectal pad cells.

When the faeces passes through the rectum, the aggregation pheromone is absorbed in the faeces and then is excreted with the faeces, which can induce all individuals including nymphs and adults aggregate (Ishii and Kuwahara, 1970; Nojima et al., 2005). In addition, some symbionts especially gut microbes play an important role in the synthesis of pheromones. Therefore, we can use appropriate technology to extract or artificially synthesize aggregation pheromones, which are combined with pesticides for cockroach control. Changing the distribution of cockroaches through push–pull strategy combined with other methods to control cockroaches not only can reduce the dosage of pesticides but also improve the efficiency of pesticides.

ACKNOWLEDGEMENTS

This study was supported by the National Natural Science Foundation of China (82072027), the Key Programs for Science and Technology Development of Shandong Province (2017GSF221012) and the Shandong Province Medical and Health Science and Technology Development Program (2014WS0386).

REFERENCES

Abraham, N.M., 2017. Pathogen-mediated manipulation of arthropod microbiota to promote infection. Proc. Natl. Acad. Sci. U. S. A. 114 (5), E781–E790.

Ahmed, M.Z., 2015. The intracellular bacterium wolbachia uses parasitoid wasps as phoretic vectors for efficient horizontal transmission. PLoS Pathog. 10 (2), e1004672.

Akbar, N., Siddiqui, R., Iqbal, M., Sagathevan, K., Khan, N.A., 2018. Gut bacteria of cockroaches are a potential source of antibacterial compound(s). Lett. Appl. Microbiol. 66 (5), 416–426.

Akinjogunla, O.J., Odeyemi, A.T., Udoinyang, E.T., 2012. Cockroaches (*Periplaneta americana* and *Blattella germanica*.): reservoirs of multi drug resistant (MDR) bacteria in Uyo, Akwa Ibom State. Sci. J. Biol. Sci. 1 (2), 269–279.

Arora, A.K., Douglas, A.E., 2017. Hype or opportunity? Using microbial symbionts in novel strategies for insect pest control. J. Insect Physiol. 103, 10–17.

Ayayee, P.A., Keeney, G., Sabree, Z.L., 2017. Compositional differences among female-associated and embryo-associated microbiota of the viviparous Pacific beetle cockroach, Diploptera punctata. FEMS Microbiol. Ecol. 93 (6).

Backhed, F., Ley, R.E., Sonnenburg, J.L., Peterson, D.A., Gordon, J.I., 2005. Host-bacterial mutualism in the human intestine. Science 307, 1915–1920.

Bell, W.J., Adiyodi, K.G., 1981. The American cockroach. Q. Rev. Biol.

Berlanga, M., 2015. Functional symbiosis and communication in microbial ecosystems. The case of wood-eating termites and cockroaches. Int. Microbiol. 18, 159–169.

Berlanga, M., 2016. Gut bacterial community of the xylophagous cockroach cryptocercus punctulatus and parasphaeria boleiriana. PLoS One 11 (4), e0152400.

Bertino-Grimaldi, D., et al., 2013. Bacterial community composition shifts in the gut of *periplaneta americana* fed on different lignocellulosic materials. SpringerPlus 2 (1), 609.

Bourtzis, K., et al., 2014. Harnessing mosquito-Wolbachia symbiosis for vector and disease control. Acta Trop. 132, 150–163.

Bright, M., Bulgheresi, S., 2010. A complex journey: transmission of microbial symbionts. Nat. Rev. Microbiol. 8 (3), 218–230.

Brumin, M., Kontsedalov, S., Ghanim, M., 2011. Rickettsia influences thermotolerance in the whitefly Bemisia tabaci B biotype. Insect Sci. 18 (1), 57–66.

Brune, A., 2014. Symbiotic digestion of lignocellulose in termite guts. Nat. Rev. Microbiol. 12, 168–180.

Brune, A., Dietrich, C., 2015. The gut microbiota of termites: digesting the diversity in the light of ecology and evolution. Annu. Rev. Microbiol. 69, 145–166.

Brune, A., Ohkuma, M., 2011. Role of the termite gut microbiota in symbiotic digestion. In: Bignell, D.E., Roisin, Y., Lo, N. (Eds.), Biology of Termites: A Modern Synthesis. Springer, Dordrecht, pp. 43–75.

Burke, G.R., Strand, M.R., 2012. Polydnaviruses of parasitic wasps: domestication of viruses to act as gene delivery vectors. Insects 3 (1), 91–119.

Burke, G.R., et al., 2010. Dynamics of a recurrent Buchnera mutation that affects thermal tolerance of pea aphid hosts. Genetics 186 (1), 367–372.

Caccia, S., Di, L.I., La, S.A., Marinelli, A., Varricchio, P., Franzetti, E., et al., 2016. Midgut microbiota and host immunocompetence underlie bacillus thuringiensis killing mechanism. Proc. Natl. Acad. Sci. U. S. A. 113 (34), 9486.

Carrasco, P., 2014. Succession of the gut microbiota in the cockroach *Blattella germanica*. Int. Microbiol. 17 (2), 99.

Chandrashekar, M.R., Rathish, K.C., Nagesha, C.N., 1997. Reservoirs of nosocomial pathogens in neonatal intensive care unit. J. Indian Med. Assoc. 95 (3), 72–74.

Chaput, C., Boneca, I.G., 2007. Peptidoglycan detection by mammals and flies. Microbes Infect. 9 (5), 637–647.

Cheng, D., et al., 2017. Gut symbiont enhances insecticide resistance in a significant pest, the oriental fruit fly bactrocera dorsalis, (hendel). Microbiome 5 (1), 13.

Christian, P.D., 2005. In: Fauquet, C.M., et al., (Eds.), Dicistroviridae. Virus Taxonomy: Eighth Report of the International Committee on the Taxonomy of Viruses. Elsevier Academic Press, pp. 783–788.

Clark, J.W., Hossain, S., Burnside, C.A., Kambhampati, S., 2001. Coevolution between a cockroach and its bacterial endosymbiont: a biogeographical perspective. Proc. Biol. Sci. 268 (1465), 393–398.

Corpus, L.D., Sikorowski, P.P., 1992. Susceptibility of *Periplaneta brunnea* (Dictyoptera: Blattidae) to the nematode Steinernema carpocapsae (Rhabditida: Steinernematidae). J. Med. Entomol. 29, 707–710.

Costerton, J.W., 1992. Pivotal role of biofilms in the focused attack of bacteria on insoluble substrates. Int. Biodeter. Biodegr. 30, 123–133.

Costerton, J.W., Lewandowski, Z., Caldwell, D.E., Korber, D.R., Lappinscott, H.M., 1995. Microbial biofilms. Annu. Rev. Microbiol. 49 (1), 711–745.

Cui, X.H., Wan, F.H., Xie, M., Liu, T.X., 2008. Effects of heat shock on survival and reproduction of two whitefly species, Trialeurodes vaporariorum and Bemisia tabaci biotype B. J. Insect Sci. 8, 24.

Daily, J.P., 2012. Malaria vaccine trials beyond efficacy end points. N. Engl. J. Med. 367 (24), 2349–2351.

Dale, C., Moran, N.A., 2006. Molecular interactions between bacterial symbionts and their hosts. Cell 126, 453–465.

Dedeine, F., Vavre, F., Fleury, F., Loppin, B., Hochberg, M.E., Bouletreau, M., 2001. Removing symbiotic *Wolbachia* bacteria specifically inhibits oogenesis in a parasitic wasp. Proc. Natl. Acad. Sci. U. S. A. 98 (11), 6247–6252.

Dillon, R.J., Vennard, C.T., Charnley, A.K., 2000. Pheromones-exploitation of gut bacteria in the locust. Nature 403 (6772), 851.

Do, D.C., Zhao, Y., Gao, P., 2016. Cockroach allergen exposure and risk of asthma. Allergy 71 (4), 463–474.

Dong, P., Wang, J.J., 2006. Reproductive manipulation of Wolbachia to its hosts. Chin. Bull. Entomol. 43 (3), 288–293.

Du, M.C.S., Zhao, Q.H., Ma, G., Yang, H.P., 2007. Effects of heat stress on physiological and biochemical mechanisms of insects: a literature review. Acta Ecol. Sin. 27 (4), 1565–1572.

Duan, H.Y., 2008. Microbial aerosol and its propagation to the surrounding environment. Doctoral thesis of Shandong Agricultural University, Shandong Agricultural University.

Dumn, O., tiurst, G.D.D., Hornett, E.A., 2008. High incidence of the matemally inherited bacterium Cardinium in spiders. Mol. Ecol. 17 (6), 1427–1437.

Dunbar, H.E., Wilson, A.C., Ferguson, N.R., Moran, N.A., 2007. Aphid thermal tolerance is governed by a point mutation in bacterial symbionts. PLoS Biol. 5 (5), e96.

Durvasula, R.V., Gumbs, A., Panackal, A., Kruglov, O., Taneja, J., Kang, A.S., Cordon-Rosales, C., Richards, F.F., Whitham, R.G., Beard, C.B., 1999. Expression of a functional antibody fragment in the gut of Rhodnius prolixus via transgenic bacterial symbiont Rhodococcus rhodnii. Med. Vet. Entomol. 13 (2), 115–119.

Engel, P., Moran, N.A., 2013. Functional and evolutionary insights into the simple yet specific gut microbiota of the honey bee from metagenomic analysis. Gut. Microbes. 4 (1), 60–65.

Feldhaar, H., 2011. Bacterial symbionts as mediators of ecologically important traits of insect hosts. Ecol. Entomol. 36 (5), 533–543.

Feldhaar, H., Gross, R., 2008. Immune reactions of insects on bacterial pathogens and mutualists. Microbes Infect. 10 (9), 1082–1088.

Flint, H.J., Bayer, E.A., Rincon, M.T., Lamed, R., White, B.A., 2008. Polysaccharide utilization by gut bacteria: potential for new insights from genomic analysis. Nat. Rev. Microbiol. 6, 121–131.

Fröhlich, J., Koustiane, C., Kämpfer, P., 2007. Occurrence of rhizobia in the gut of the higher termite Nasutitermes nigriceps. Syst. Appl. Microbiol. 30 (1), 68–74.

Fu, X., Ye, L., Ge, F., 2009. Habitat influences on diversity of bacteria found on German cockroach in Beijing. J. Environ. Sci. 21 (2), 249–254.

Gal, R., Libersat, F., 2010. On predatory wasps and zombie cockroaches. Commun. Integr. Biol. 3, 458–461.

Gao, M.J., Yao, X.Q., 2012. Research progress on pathogenic mechanism of insect pathogenic fungi. Guizhou. Agr. Sci. Tech. 40 (3), 130–132.

Gehrer, L., Vorburger, C., 2012. Parasitoids as vectors of facultative bacterial endosymbionts in aphids. Biol. Lett. 8 (4), 613–615.

Giorgi, F., Nordin, J.H., 1994. Structure of yolk granules in oocytes and eggs of *Blattella germanica* and their interaction with vitellophages and endosymbiotic bacteria during granule degradation. J. Insect Physiol. 40, 1077–1092.

Glaser, R.L., Meola, M.A., 2010. The native Wolbachia endosymbionts of Drosophila melanogaster and Culex quinquefasciatus increase host resistance to West Nile virus infection. PLoS One 5 (8), e11977.

Gong, P., Shen, Z.R., Li, Z.H., 2002. Symbiotic bacteria in Wolbachia and their regulation on arthropod reproductive activities. Acta Entomol. Sin. 45 (2), 241–252.

Gontang, E.A., Aylward, F.O., Carlos, C., et al., 2017. Major changes in microbial diversity and community composition across gut sections of a juvenile Panchlora cockroach. PLoS One 12, 1–21.

González-Domenech, C.M., Belda, E., Patiño-Navarrete, R., Moya, A., Peretó, J., Latorre, A., 2012. Metabolic stasis in an ancient symbiosis: genome-scale metabolic networks from two Blattabacterium cuenoti strains, primary endosymbionts of cockroaches. BMC Microbiol. 12 (1), S5.

Goralska, G., Kurnatowski, P., Blaszkowska, J., 2013. Occurrence of dermatophytes in the soil of recreation places inhabitants of Lodz. Ann. Parasitol. 59 (Suppl. P158).
Gore, J.C., Schal, C., 2007. Cockroach allergen biology and mitigation in the indoor environment. Annu. Rev. Entomol. 52, 439–463.
Gressel, J., 2018. Microbiome facilitated pest resistance: potential problems and uses. Pest Manag. Sci. 74, 511–515.
Gueguen, G., Vavre, F., Gnankine, O., Peterschmitt, M., Charif, D., Chiel, E., Gottlieb, Y., Ghanim, M., Zchori-Fein, E., Fleury, F., 2010. Endosymbiont metacommunities, mtDNA diversity and the evolution of the Bemisia tabaci (Hemiptera: Aleyrodidae) species complex. Mol. Ecol. 19, 4365–4378.
Gündüz, E.A., Douglas, A.E., 2009. Symbiotic bacteria enable insect to use a nutritionally inadequate diet. Proc. Biol. Sci. 276 (1658), 987–991.
Hamilton, P.T., Leong, J.S., Koop, B.F., Perlman, S.J., 2014. Transcriptional responses in a Drosophila defensive symbiosis. Mol. Ecol. 23 (6), 1558–1570.
Harumoto, T., Anbutsu, H., Fukatsu, T., 2014. Male-killing spiroplasma induces sex-specific cell death via host apoptotic pathway. PLoS Pathog. 10 (2), e1003956.
Hayashi, T., Hosokawa, T., Meng, X.Y., Koga, R., Fukatsu, T., 2015. Female-specific specialization of a posterior end region of the midgut symbiotic organ in *Plautia splendens* and *Allied stinkbugs*. Appl. Environ. Microbiol. 81 (7), 2603.
Hayatsu, M., Hirano, M., Tokuda, S., 2000. Involvement of two plasmids in fenitrothion degradation by Burkholderia sp. strain NF100. Appl. Environ. Microbiol. 66, 1737–1740.
Heal, R.E., Nash, K.B., Williams, M., 1953. An insecticide resistanced strain of the German cockraock from corpus chriati Texas. J. Econ. Entomol. 46, 385–386.
Hedges, L.M., Brownlie, J.C., O'Neill, S.L., Johnson, K.N., 2008. *Wolbachia* and virus protection in insects. Science 322, 702.
Hehemann, J.H., Correc, G., Barbeyron, T., Helbert, W., Czjzek, M., Michel, G., 2010. Transfer of carbohydrate-active enzymes from marine bacteria to Japanese gut microbiota. Nature 464 (7290), 908–912.
Herren, J.K., Lemaitre, B., 2011. Spiroplasma and host immunity: activation of humoral immune responses increases endosymbiont load and susceptibility to certain gram-negative bacterial pathogens in drosophila melanogaster. Cell. Microbiol. 13 (9), 1385.
Herzner, G., Schlecht, A., Dollhofer, V., 2013. Larvae of the parasitoid wasp Ampulex compressa sanitize their host, the American cockroach, with a blend of antimicrobials. Proc. Natl. Acad. Sci. U. S. A. 110, 1369–1374.
Hilgenboecker, K., Hammerstein, P., Schlattmann, P., Telschow, A., Werren, J.H., 2008. How many species are infeeted with *Wlbaehia*—a statistical analysis of current data. FEMS Microbiol. Lett. 281, 215–220.
Hongoh, Y., 2010. Diversity and genomes of uncultured microbial symbionts in the termite gut. Biosci. Biotechnol. Biochem. 74 (6), 1145–1151.
Huang, Y.H., Zhang, F., Fu, R.S., Liu, S.Z., Hou, X.B., 2011. Analysis of intestinal flora isolation and antifungal activity of *Blattella germanica*. China Public Health 27 (1), 78–79.
Huang, Y.H., Wang, X.J., Zhang, F., Huo, X.B., Fu, R.S., Liu, J.J., et al., 2013. The identification of a bacterial strain bgi-1 isolated from the intestinal flora of *blattella germanica*, and its anti-entomopathogenic fungi activity. J. Econ. Entomol. 106 (1), 43–49.
Hughes, A.L., 2012. Evolution of the betaGRP /GNBP /beta-1, 3-glucanase family of insects. Immunogenetics 64 (7), 549–558.
Hurst, G.D.D., Jiggins, F.M., 2000. Male-killing bacteria in insects: mechanisms, incidence, and implications. Emerg. Infect. Dis. 6 (6), 329–336.
Hurst, G.D.D., Jiggins, F.M., Majerus, M.E.N., 2003. Inherited microorganisms that selectively kill male hosts: the hidden players of insect evolution? In: Insect Symbiosis. CRC. Press, Boca Raton, FL, pp. 177–197.

Ishii, S., Kuwahara, Y., 1970. An aggregation pheromone of the German cockroach (*Blattella germanica L.*) 2. Species specificity of the pheromone. Appl. Entomol. Zool. 5 (1), 33–41.

Jaenike, J., Unckless, R., Cockburn, S.N., Boelio, L.M., Perlman, S.J., 2010. Adaptation via symbiosis: recent spread of a Drosophila defensive symbiont. Science 329 (5988), 212.

Jiang, H., 2003. Research progress of German cockroach control. Chin. Health Pestic. Mach. 9 (23), 44–45.

Jiang, H., Vilcinskas, A., Kanost, M.R., 2010. Immunity in lepidopteran insects. Adv. Exp. Med. Biol. 708, 181–204.

Jin, X.B., FJ, C., Zhu, J.Y., 2008. Molecular identification and analysis of German cockroach symbiotic bacteria *Wolbachia*. Chin. J. Vector. Biol. Control 19, 121–122.

Jiravanichpaisal, P., Lee, B.L., Söderhäll, K., 2006. Cell-mediated immunity in arthropods: hematopoiesis, coagulation, melanization and opsonization. Immunobiology 211 (4), 213–236.

Jones, A.M., Govan, J.R.W., Doherty, C.J., et al., 2003. Identification of airborne dissemination of epidemic multiresistant strains of *Pseudomonas* aeruginosa at a CF centre during a cross infection outbreak. Thorax 58, 525–527.

Kaiwa, N., Hosokawa, T., Nikoh, N., Tanahashi, M., Moriyama, M., Meng, X.Y., 2014. Symbiont-supplemented maternal investment underpinning host's ecological adaptation. Curr. Biol. 24 (20), 2465.

Kaltenpoth, M., 2009. Actinobacteria as mutualists: general healthcare for insects? Trends Microbiol. 17 (12), 529–535.

Kaltenpoth, M., Gottler, W., Herzner, G., Strohm, E., 2005. Symbiotic bacteria protect wasp larvae from fungal infestation. Curr. Biol. 15, 475–479.

Kellner, R.L.L., 2002. Molecular identification of an endosymbiotic bacterium associated with pederin biosynthesis in paederus sabaeus (coleoptera: staphylinidae). Insect Biochem. Mol. Biol. 32 (4), 389.

Kellner, R.L.L., Dettner, K., 1996. Differential efficacy of toxic pederin in deterring potential arthropod predators of paederus (coleoptera: staphylinidae) offspring. Oecologia 107 (3), 293–300.

Kikuchi, Y., Meng, X.Y., Fukatsu, T., 2005. Gut symbiotic bacteria of the genus Burkholderia in the broad-headed bugs *Riptortus clavatus* and *Leptocorisa chinensis* (Heteroptera: Alydidae). Appl. Environ. Microbiol. 71, 4035–4043.

Kikuchi, Y., Hosokawa, T., Fukatsu, T., 2011a. Specific developmental window for establishment of an insect-microbe gut symbiosis. Appl. Environ. Microbiol. 77, 4075–4081.

Kikuchi, Y., Hosokawa, T., Fukatsu, T., 2011b. An ancient but promiscuous host–symbiont association between Burkholderia gut symbionts and their heteropteran hosts. ISME J. 5, 446–460.

Kikuchi, Y., Hayatsu, M., Hosokawa, T., Nagayama, A., Tago, K., Fukatsu, T., 2012. Symbiont-mediated insecticide resistance. Proc. Natl. Acad. Sci. U. S. A. 109, 8618–8622.

Kleino, A., Myllymäki, H., Kallio, J., Vanhaaho, L.M., Oksanen, K., Ulvila, J., et al., 2008. Pirk is a negative regulator of the drosophila imd pathway. J. Immunol. 180 (8), 5413.

Koehler, P.G., Patterson, R.S., Martin, W.R., 1992. Susceptibility of cockroaches (dictyoptera: blattellidae, blattidae) to infection by steinernema carpocapsae. J. Econ. Entomol. 85 (4), 1184.

Kopanic, R.J., Holbrook, G.L., Sevala, V., Schal, C., 2001. An adaptive benefit of facultative coprophagy in the German cockroach *Blattella germanica*. Ecol. Entomol. 26 (2), 154–162.

Khachatryan, Z.A., Ktsoyan, Z.A., Manukyan, G.P., Kelly, D., Ghazaryan, K.A., Aminov, R.I., 2008. Predominant role of host genetics in controlling the composition of gut microbiota. PLoS One 3 (8), e3064.

Kroiss, J., Kaltenpoth, M., Schneider, B., Schwinger, M.G., Hertweck, C., Maddula, R.K., et al., 2010. Symbiotic streptomycetes provide antibiotic combination prophylaxis for wasp offspring. Nat. Chem. Biol. 6 (4), 261.

Kuechler, S.M., Yu, M., Dettner, K., Kikuchi, Y., 2016. Phylogenetically diverse burkholderia associated with midgut crypts of spurge bugs, dicranocephalus spp. (heteroptera: stenocephalidae). Microbes Environ. 31 (2), 145–153.

Kulhankova, K., George, C.L., Kline, J.N., Snyder, J.M., Darling, M., Field, E.H., Thorne, P.S., 2009. Early-life co-administration of cockroach allergen and endotoxin augments pulmonary and systemic responses. Clin. Exp. Allergy 39 (7), 1069–1079.

Kulhankova, K., George, C.L., Kline, J.N., Darling, M., Thorne, P.S., 2012. Endotoxin inhalation alters lung development in neonatal mice. Am. J. Ind. Med. 55 (12), 1146–1158.

Lai, K.M., 2017. Are cockroaches an important source of indoor endotoxins? Int. J. Environ. Res. Public Health 14 (1), 91.

Laven, H., 1967. Speciation and evolution in Culex pipiens. In: Wright, J., Pal, R. (Eds.), Genetics of Insect Vectors of Disease. Elsevier, Amsterdam, Holland, pp. 251–275.

Lee, K.A., Kim, S.H., Kim, E.K., Ha, E.M., You, H., Kim, B., et al., 2013. Bacterial-derived uracil as a modulator of mucosal immunity and gut–microbe homeostasis in Drosophila. Cell 153 (4), 797–811.

Leonard, J., Cordoba-Aquilar, A., 2010. The Evolution of Primary Sexual Characters in Animals. Oxford University Press, Oxford.

Li, N., Chen, J.M., Zhang, Y.F., YP, H., Chen, L.Z., 2010. Comparison of activities of detoxifying enzymes between imidacloprid resistant strains and susceptible strains of Nilaparvata lugens. Acta Agric. Zhejiangensis 22, 653–659.

Li, L.W., Zhang, X.C., Xia, J., Zhang, S., Zhao, D.Q., Zhang, F., 2016. Removal of symbiotic bacteria on the biological fitness of Blattella germanica. J. Shandong Norm. Univ. 31 (4), 143–147.

Li, H., Zhang, F.M., Guo, H.Y., Zhu, Y.Y., Yuan, J.D., Yang, G.W., An, L.G., 2013. Molecular characterization of hepcidin gene in common carp (Cyprinus carpio L.) and its expression pattern responding to bacterial challenge. Fish. Shellfish. Immun. 35 (3), 1030–1038.

Li, T., Li, H., Peng, S.Q., Zhang, F.M., An, L.G., Yang, G.W., 2017. Molecular characterization and expression pattern of X box-binding protein-1 (XBP1) in common carp (Cyprinus carpio L.): Indications for a role of XBP1 in antibacterial and antiviral immunity. Fish.Shellfish. Immun. 67, 667–674.

Li, S., Zhu, S., Jia, Q., Yuan, D., Ren, C., Li, K., et al., 2018. The genomic and functional landscapes of developmental plasticity in the American cockroach. Nat. Commun. 9 (1), 1008.

Li, L., Yang, H.J., Liu, D.C., He, H.B., Wang, C.F., Zhong, J.F., Zeng, Y.J., 2012. Analysis of biofilms formation and associated genes detection in Staphylococcus isolates from bovine mastitis. Int. J. App. Res. Vet. M. 10 (1), 62–68.

Liang, J.W., Tian, F.L., Lan, Z.R., Huang, B., Zhuang, W.Z., 2010. Selection characterization on overlapping reading frame of multiple-protein-encoding P gene in Newcastle disease virus. Vet. Microbiol. 144 (3–4), 257–263.

Liu, Q.M., Hu, Z.M., Qiu, G.X., 2012. Research progress of concentrated nuclear virus (Densovirus). Sericult. Sci. 38 (1), 0165–0171.

Liu, Q., Xia, G.S., 2005. Research progress on biological control of diamondback moth. Anhui. Agri. Sci. 33 (10), 1919–1920.

Libbrecht, R., Gwynn, D.M., Mde, F., 2007. Aphidius ervi preferentially attacks the green morph of the pea aphid, acyrthosiphon pisum. J. Insect Behav. 20 (1), 25–32.

Lo, N., Bandi, C., Watanabe, H., Nalepa, C., Beninati, T., 2003. Evidence for cocladogenesis between diverse dictyopteran lineages and their intracellular endosymbionts. Mol. Biol. Evol. 20, 907–913.

López-Sánchez, M.J., Neef, A., Peretó, J., et al., 2009. Evolutionary convergence and nitrogen metabolism in Blattabacterium strain Bge, primary endosymbiont of the cockroach *Blattella germanica*. PLoS Genet. 5 (11), e1000721.
Losey, J.E., Harmon, J., Ballantyne, F., Brown, C., 1997. A polymorphism maintained by opposite patterns of parasitism and predation. Nature 388, 269–272.
Lu, Y.H., 2016. Distribution, transmission and function of *Rickettsia* in *Eretmocerus hayati*. Master thesis of Zhejiang University, Zhejiang University.
Lü, Z.C., Wan, F.H., 2008. Differential gene expression in whitefly (Bemisia tabaci) B-biotype females and males under heat-shock condition. Comp. Biochem. Physiol. D: Genomics Proteomics 3 (4), 257–262.
Luan, J.B., Shan, H.W., Philipp, I., Huang, J.H., Jan, L., Liu, S.S., 2016. Cellular and molecular remodelling of a host cell for vertical transmission of bacterial symbionts. Proc. Biol. Sci. 283 (1833), 20160580.
Łukasik, P., Dawid, M.A., Ferrari, J., Godfray, H.C., 2013a. The diversity and fitness effects of infection with facultative endosymbionts in the grain aphid, sitobion avenae. Oecologia 173 (3), 985–996.
Łukasik, P., Asch, M.V., Guo, H., Ferrari, J., Godfray, H.C.J., 2013b. Unrelated facultative endosymbionts protect aphids against a fungal pathogen. Ecol. Lett. 16 (2), 214–218.
Ma, W.J., Zhang, W., Wei, J.F., Zhang, H., Liu, C.C., He, S.H., 2011. Production and identification of monoclonal antibodies against cockroach allergen Bla g7. J. Allergy. Clin. Immun. 5 (4), 253–257.
Mahadav, A., Kontsedalov, S., Czosnek, H., Ghanim, M., 2009. Thermotolerance and gene expression following heat stress in the whitefly *Bemisia tabaci* B and Q biotypes. Insect Biochem. Mol. Biol. 39 (10), 668–676.
Martã-Nez-Girã, N.R., Martã-Nez-Torre, C., van Woerden, H.C., 2017. The prevalence of protozoa in the gut of german cockroaches (*Blattella germanica*) with special reference to *Lophomonas blattarum*. Parasitol. Res. 116 (11), 3205–3210.
Martinari, L., Varcamonti, M., Naclerio, G., De, F.M., 2002. Purification and partial characterization of bacillocin 490, a novel bacteriocin produced by a thermophilic strain of *Bacillus licheniformis*. Microb. Cell Fact. 1 (1), 1–5.
Marroquin, L.D., Elyassnia, D., Griffitts, J.S., Feitelson, J.S., Aroian, R.V., 2000. *Bacillus thuringiensis* (bt) toxin susceptibility and isolation of resistance mutants in the nematode Caenorhabditis elegans. Genetics 155 (4), 1693–1699.
Mathieu, L.G., Sonea, S., 1996. Review of the unique mode of evolution of bacteria: an opinion. Symbiosis 21 (3), 199–207.
Mcgraw, E.A., O'Neill, S.L., 2013. Beyond insecticides: new thinking on an ancient problem. Nat. Rev. Microbiol. 11 (3), 181–193.
Mcmeniman, C.J., Lane, R.V., Cass, B.N., Fong, A.W., Sidhu, M., Wang, Y.F., et al., 2009. Stable introduction of a life-shortening wolbachia infection into the mosquito aedes aegypti. Science 323 (5910), 141–144.
Mikaelyan, A., Thompson, C.L., Hofer, M.J., Brune, A., 2015. The deterministic assembly of complex bacterial communities in germ-free cockroach guts. Appl. Environ. Microbiol. 82 (4), 1256–1263.
Montllor, C.B., Maxmen, A., Purcell, A.H., 2002. Facultative bacterial endosymbionts benefit pea aphids Acyrthosiphon pisum under heat stress. Ecol. Entomol. 27 (2), 189–195.
Moreau, S.J.M., Asgari, S., 2015. Venom proteins from parasitoid wasps and their biological functions. Toxins 7 (7), 2385–2412.
Mueller, U.G., Mikheyev, A.S., Hong, E., Sen, R., Warren, D.L., Solomon, S.E., et al., 2011. Evolution of cold-tolerant fungal symbionts permits winter fungiculture by leafcutter ants at the northern frontier of a tropical ant-fungus symbiosis. Proc. Natl. Acad. Sci. U. S. A. 108 (10), 4053–4056.
Müller, F.P., 1962. Biotypen und unterarten der "erbsenlaus" acyrthosiphon pisum (harris). Z. Pflanzenkr. Pflanzenschutz 69 (3), 129–136.

Myllymäki, H., Rämet, M., 2014. Jak/stat pathway in drosophila immunity. Scand. J. Immunol. 79 (6), 377–385.
Nalepa, C.A., 2015. Origin of termite eusociality: trophallaxis integrates the social, nutritional, and microbial environments. Ecol. Entomol. 40 (4), 323–335.
Nalepa, C.A., 2017. What kills the hindgut flagellates of lower termites during the host molting cycle? Microorganisms 5 (4), 82.
Nalepa, C.A., Bignell, D.E., Bandi, C., 2001. Detritivory, coprophagy, and the evolution of digestive mutualisms in Dictyoptera. Insect. Soc. 48 (3), 194–201.
Nancy, B., Damián, G.N., Julia, B., Sheehan, M.G., Arias, L.M.P., Melina, R., 2008. Prescription and use of antibiotics at a pediatric intensive care unit in Buenos Aires City. Arch. Argent. Pediatr. 106 (5), 409–415.
Naqqash, M.N., Gökçe, A., Bakhsh, A., Salim, M., 2016. Insecticide resistance and its molecular basis in urban insect pests. Parasitol. Res. 115 (4), 1363–1373.
Natarajan, S., Kim, J., Bouchard, J., Cruikshank, W., Remick, D.G., 2011. Reducing lps content in cockroach allergens increases pulmonary cytokine production without increasing inflammation: a randomized laboratory study. BMC Pulm. Med. 11 (1), 12.
Nishiwaki, H., Nakashima, K., Ishida, C., Kawamura, T., Matsuda, K., 2007. Cloning, functional characterization, and mode of action of a novel insecticidal pore-forming toxin, sphaericolysin, produced by Bacillus sphaericus. Appl. Environ. Microbiol. 73 (10), 3404–3411.
Nojima, J., Kuratsune, J., Suehisa, E., 2005. Acquired activated protein C resistance associated with IgG antibodies against 2-glycoprotein I and prothrombin as a strong risk factor for venous thromboembolism. Clin. Chem. 51, 545–552.
Oh, S.N., Mi, J.S., Youn, Y.N., Yu, Y.M., 2015. Antifungfal activity against plant pathogenic fungi on insect enterobacteriaceae. Korean J. Pestic. Sci. 19 (1), 71–79.
Ohkuma, M., Noda, S., Hongoh, Y., Nalepa, C.A., Inoue, T., 2009. Inheritance and diversification of symbiotic trichonymphid flagellates from a common ancestor of termites and the cockroach cryptocercus. Proc. Biol. Sci. 276 (1655), 239–245.
Oliva, G.R., Díaz, C., González, O.F., Martínez, M.D., Fernández, C., Cordoví, R., 2001. *Blattella germanica* as a possible cockroach vector of micro-organisms in a hospital. J. Hosp. Infect. 74 (1), 93–95.
Oliver, K.M., Russell, J.A., Moran, N.A., Hunter, M.S., 2003. Facultative bacterial symbionts in aphids confer resistance to parasitic wasps. Proc. Natl. Acad. Sci. U. S. A. 100 (4), 1803–1807.
Oliver, K.M., Degnan, P.H., Hunter, M.S., Moran, N.A., 2009. Bacteriophages encode factors required for protection in a symbiotic mutualism. Science 325, 992–994.
Oliver, K.M., Degnan, P.H., Burke, G.R., Moran, N.A., 2010. Facultative symbionts in aphids and the horizontal transfer of ecologically important traits. Annu. Rev. Entomol. 55, 247–266.
Ozdal, M., Ozdal, O.G., Algur, O.F., Kurbanoglu, E.B., 2017. Biodegradation of α-endosulfan via hydrolysis pathway by Stenotrophomonas maltophilia, og2. 3 Biotech 7 (2), 113.
Palleroni, N.J., 1984. Pseudomonadaceae. In: Krieg, N.R., Holt, J.G. (Eds.), Bergey's Manual of Systematic Bacteriology. Williams and Wilkins, Baltimore, pp. 141–199.
Pan, H.E., Qiang, M.A., 2017. Research progresson harm distribution species and control of sanitary pest cockroach in China. China Health Care Nutr. 23, 62–64.
Pan, X.L., Liu, Q.Y., Xi, Z.Y., 2014. Advance in developing *Wolbachia* as a mean to control mosquito and mosquito-borne diseases. Chin. J. Vector. Biol. Control 25 (1), 1–7.
Pannebakker, B.A., Pijnacker, L.P., Zwaan, B.J., 2004. Cytology of Wolbachia-induced parthenogenesis in Leptopilina clavipes (Hymenoptera: Figitidae). Genome 47 (2), 299.

Pardo, S., Martínez, A.M., Figueroa, J.I., Chavarrieta, J.M., Viñuela, E., 2017. Insecticide resistance of adults and nymphs of Asian citrus psyllid populations from Apatzingán Valley, Mexico. Chem. Geol. 412 (1), 34–47.
Patiño-Navarrete, R., Piulachs, M.D., Belles, X., Moya, A., Latorre, A., Peretó, J., 2014. The cockroach *blattella germanica* obtains nitrogen from uric acid through a metabolic pathway shared with its bacterial endosymbiont. Biol. Lett. 10 (7), 384–396.
Pérezcobas, A.E., Maiques, E., Angelova, A., Carrasco, P., Moya, A., Latorre, A., 2015. Diet shapes the gut microbiota of the omnivorous cockroach *Blattella germanica*. FEMS Microbiol. Ecol. 91 (4), 1–14.
Pomés, A., 2008. Cockroach and other inhalant insect allergens. Clin. Allergy Immunol. 21 (18), 183.
Pomés, A., Arruda, L.K., 2014. Investigating cockroach allergens: aiming to improve diagnosis and treatment of cockroach allergic patients. Methods 66 (1), 75–85.
Pomes, A., Mueller, G.A., Randall, T.A., Chapman, M.D., Arruda, L.K., 2017. New insights into cockroach allergens. Curr. Allergy Asthma Rep. 17 (4), 1–16.
Prazmo, Z., Dutkiewicz, J., Skorska, C., et al., 2003. Exposure to airborne gram-negative bacteria, dust, and endotoxin in paper factories. Ann. Agric. Environ. Med. 10 (1), 93–100.
Radek, R., 2000. Light and electron microscopic study of a Rickettsiella species from the cockroach Blatta orientalis. J. Invertebr. Pathol. 76 (4), 249–256.
Rao, Q., Wang, S., Su, Y.L., et al., 2012. Draft genome sequence of "Candidatus Hamiltonella defensa," an endosymbiont of the whitefly Bemisia tabaci. J. Bacteriol. 194 (13), 3558.
Rasmussen, R.A., Khalil, M.A.K., 1983. Global production of methane by termites. Nature 301 (5902), 700–702.
Riparbelli, M.G., Giordano, R., Ueyama, M., Callaini, G., 2012. *Wolbachia*-mediated male killing is associated with defective chromatin remodeling. PLoS One 7 (1), e30045.
Rivault, C., Cloarec, A., Le Guyader, A., 1993. Bacterial load of cockroaches in relation to urban environment. Epidemiol. Infect. 110 (2), 317–325.
Ros, V.I.D., Fleming, V.M., Feil, E.J., et al., 2009. How diverse is the genus *Wolbachia*? Multiple-gene sequencing reveals a putatively new *Wolbachia* supergroup recovered from spider mites (Acari: Tetranychidae). Appl. Environ. Microbiol. 75 (4), 1036–1043.
Rosas, T., Garcíaferris, C., Domínguezsantos, R., Llop, P., Latorre, A., Moya, A., 2018. Rifampicin treatment of *Blattella germanica* evidences a fecal transmission route of their gut microbiota. FEMS Microbiol. Ecol. 94 (2).
Royet, J., Dziarski, R., 2007. Peptidoglycan recognition proteins: pleiotropic sensors and effectors of antimicrobial defences. Nat. Rev. Microbiol. 5 (4), 264–277.
Ruan, C.L., Mi, Z., Zhu, Y., 2012. Research progress on mechanism of insect resistance to insecticides. Sci. Sericulture 38 (2), 322–328.
Rust, M.K., Owens, J.M., Reierson, D.A., 1995. Understanding and Controlling the German Cockroach. Oxford University Press, New York.
Ruxton, G.D., Sherratt, T.N., Speed, M.P., 2004. Avoiding Attack: The Evolutionary Ecology of Crypsis, Warningsignals, and Mimicry. Oxford University Press.
Rombout, J., Yang, G.W., Kiron, V., 2014. Adaptive immune responses at mucosal surfaces of teleost fish. Fish. Shellfish. Immun. 40 (2), 634–643.
Ryu, J.H., Kim, S.H., Lee, H.Y., Jin, Y.B., Nam, Y.D., Bae, J.W., et al., 2008. Innate immune homeostasis by the homeobox gene caudal and commensal-gut mutualism in drosophila. Science 319 (5864), 777–782.
Sabate, D.C., Carrillo LAudisio, M.C., 2009. Inhibition of Paenibacillus larvae and Ascosphaera apis by bacillus subtilis isolated from honeybee gut and honey samples. Res. Microbiol. 160 (3), 193–199.

Sabree, Z.L., Kambhampati, S., Moran, N.A., 2009. Nitrogen recycling and nutritional provisioning by *Blattabacterium*, the cockroach endosymbiont. Proc. Natl. Acad. Sci. 106 (46), 19521–19526.

Sacchi, L., Corona, S., Grigolo, A., et al., 1996. The fate of the endocytobionts of *Blattella germanica* (Blattaria: Blattellidae) and *Periplaneta Americana* (Blattaria: Blattidae) during embryo development. Ital. J. Zool. 63, 1–11.

Salvucci, M.E., Stecher, D.S., Henneberry, T.J., 2000. Heat shock proteins in whiteflies, an insect that accumulates sorbitol in response to heat stress. J. Therm. Biol. 25 (5), 363–371.

Scarborough, C.L., Ferrari, J., Godfray, H.C., 2005. Aphid protected from pathogen by endosymbiont. Science 310 (5755), 1781.

Schaefer, C.W., Panizzi, A.R., 2000. Heteroptera of Economic Importance. CRC, Boca Raton, FL.

Schauer, C., Thompson, C.L., Brune, A., 2012. The bacterial community in the gut of the cockroach Shelfordella lateralis reflects the close evolutionary relatedness of cockroaches and termites. Appl. Environ. Microbiol. 78, 2758–2767.

Schauer, C., Thompson, C., Brune, A., 2014. Pyrotag sequencing of the gut microbiota of the cockroach shelfordella lateralis reveals a highly dynamic core but only limited effects of diet on community structure. PLoS One 9 (1), e85861.

Sen, R., Ishak, H.D., Estrada, D., Dowd, S.E., Hong, E., Mueller, U.G., 2009. Generalized antifungal activity and 454-screening of Pseudonocardia and Amycolatopsis bacteria in nests of fungus-growing ants. Proc. Natl. Acad. Sci. U. S. A. 106, 17805–17810.

Shan, H.W., Liu, S.S., Liu, Y.Q., 2016. Responses of insect holobionts to unfavorable temperatures. J. Plant Prot. 43 (1), 24–31.

Shan, S.J., Liu, D.Z., Liu, R.R., Zhu, Y.Y., Li, T., Zhang, F.M., Li, H., 2018. Non-mammalian Toll-like receptor 18 (Tlr18) recognizes bacterial pathogens in common carp (Cyprinus carpio L.): Indications for a role of participation in the NF-kappa B signaling pathway. Fish. Shellfish. Immun. 72, 187–198.

Shapiro, J.A., 1997. Multicellularity: the rule not the exception. In: Shapiro, J.A., Dworkin, M. (Eds.), Bacteria as Multicellular Organisms. Oxford University Press, New York, pp. 14–49.

Sharon, G., Segal, D., Ringo, J.M., 2010. Commensal bacteria play a role in mating preference of *Drosophila melanogaster*. Proc. Natl. Acad. Sci. U. S. A. 107 (46), 20051–20056.

Shi, Y.J., Cheng, J.A., Zhu, Z.R., et al., 2002. Research progress of insect symbiotic bacteria *Wolbachia*. J. Ecol. 22 (3), 409–419.

Shi, W., Guo, Y., Xu, C., 2014. Unveiling the mechanism by which microsporidian parasites prevent locust swarm behavior. Proc. Natl. Acad. Sci. U. S. A. 111 (4), 1343–1348.

Shokal, U., Yadav, S., Atri, J., Accetta, J., Kenney, E., Banks, K., et al., 2016. Effects of co-occurring wolbachia, and spiroplasma, endosymbionts on the drosophila, immune response against insect pathogenic and non-pathogenic bacteria. BMC Microbiol. 16, 16.

Simon, J.C., Carré, S., Boutin, M., Prunier-Leterme, N., Sabater-Muñoz, B., Latorre, A., 2003. Host-based divergence in populations of the pea aphid: insights from nuclear markers and the prevalence of facultative symbionts. Proc. Biol. Sci. 270 (1525), 1703–1712.

Singh, B.K., 2009. Organophosphorus-degrading bacteria: ecology and industrial applications. Nat. Rev. Microbiol. 7, 156–164.

Sousa, T., Paterson, R., Moore, V., Carlsson, A., Abrahamsson, B., Basit, A.W., 2008. The gastrointestinal microbiota as a site for the biotransformation of drugs. Int. J. Pharm. 363, 1–25.

Stiles, M.E., Holzapfel, W.H., 1997. Lactic acid bacteria of foods and their current taxonomy. Int. J. Exp. Pathol. 36 (1), 1–29.

Stouthamer, R., Breeuwer, J.A.J., Luck, R.F., Werren, J.H., 1993. Molecular identification of microorganisms associated with parthenogenesis. Nature 361, 66–68.
Stouthamer, R., Breeuwer, J.A., Hurst, G.D., 1999. *Wolbachia* pipientis: microbial manipulator of arthropod reproduction. Annu. Rev. Microbiol. 53, 71–102.
Swift, M.J., Heal, O.W., Anderson, J.M., 1979. Decomposition in terrestrial ecosystems. Stud. Ecol. 5 (14), 2772–2774.
Tabashnik, B.E., Mota-Sanchez, D., Whalon, M.E., Hollingworth, R.M., Carrière, Y., 2014. Defining terms for proactive management of resistance to Bt crops and pesticides. J. Econ. Entomol. 107 (2), 496–507.
Tago, K., et al., 2006. Diversity of fenitrothion-degrading bacteria in soils from distant geographical areas. Microbes Environ. 21, 58–64.
Tai, V., Carpenter, K.J., Weber, P.K., Nalepa, C.A., Perlman, S.J., Keeling, P.J., 2016. Genome evolution and nitrogen-fixation in bacterial ectosymbionts of a protist inhabiting wood-feeding cockroaches. Appl. Environ. Microbiol. 82 (15), 4682–4695. AEM.0611-16.
Tanaka, A., 2010. General accounts on the oocyte growth and the identification of vitellogenin by means of immunospecificity in the cockroach, *blattella germanica* (l.). Dev. Growth Differ. 15 (3), 153–168.
Tegtmeier, D., Thompson, C.L., Schauer, C., Brune, A., 2015. Oxygen affects gut bacterial colonization and metabolic activities in a gnotobiotic cockroach model. Appl. Environ. Microbiol. 82 (4), 1080–1089.
Teixeira, L., Ferreira, A., Ashburner, M., 2008. The bacterial symbiont Wolbachia induces resistance to RNA viral infections in *Drosophila melanogaster*. PLoS Biol. 6, 2753–2763.
Thaochan, N., Drew, R.A.I., Hughes, J.M., Vijaysegaran, S., Chinajariyawong, A., 2010. Alimentary tract bacteria isolated and identified with api-20e and molecular cloning techniques from Australian tropical fruit flies, bactrocera cacuminata and b. tryoni. J. Insect Sci. 10 (131), 131.
Thorne, P.S., Cohn, R.D., Mav, D., Arbes, S.J., Zeldin, D.C., 2009. Predictors of endotoxin levels in US housing. Environ. Health Perspect. 117 (5), 763–771.
Tinker, K.A., Ottesen, E.A., 2016. The core gut microbiome of the American cockroach, *Periplaneta americana*, is stable and resilient to dietary shifts. Appl. Environ. Microbiol. 82 (22), 6603–6610.
Tokuda, G., Elbourne, L.D.H., Kinjo, Y., Saitoh, S., Sabree, Z., Hojo, M., et al., 2013. Maintenance of essential amino acid synthesis pathways in the *blattabacterium* cuenoti symbiont of a wood-feeding cockroach. Biol. Lett. 9 (3), 176–188.
Tungtrongchitr, A., Sookrung, N., Chaicumpa, W., Indrawattana, N., Meechan, T., Thavara, U., Visitsunthorn, N., Bunnag, C., 2012. Comparison of allergenic components and biopotency in whole body extracts of wild and laboratory reared American cockroaches, *Periplaneta americana*. Asian Pac. J. Allergy Immunol. 30 (3), 231–238.
Uzsák, A., Schal, C., 2012. Differential physiological responses of the German cockroach to social interactions during the ovarian cycle. J. Exp. Biol. 215 (Pt. 17), 3037–3044.
Uzsak, A., Schal, C., 2013. Social interaction facilitates reproduction in male German cockroaches, *Blatttella germanica*. Anim. Behav. 85 (6), 1501–1509.
Vaishampayan, P.A., Dhotre, D.P., Gupta, R.P., Lalwani, P., Ghate, H., Patolea, M.S., et al., 2007. Molecular evidence and phylogenetic affiliations of *Wolbachia* in cockroaches. Mol. Phylogenet. Evol. 58, 133–145.
Vazirianzadeh, B., Dehghani, R., Mehdinejad, M., Sharififard, M., Nasirabadi, N., 2014. The first report of drug resistant bacteria isolated from the brown-banded cockroach, *Supella longipalpa*, inahvaz, South-Western Iran. J. Arthropod Borne Dis. 8 (1), 53–59.
Vorburger, C., Gehrer, L., Rodriguez, P., 2010. A strain of the bacterial symbiont Regiella insecticola protects aphids against parasitoids. Biol. Lett. 6, 109–111.

Wadakatsumata, A., Zurek, L., Nalyanya, G., Roelofs, W.L., Zhang, A., Schal, C., 2015. Gut bacteria mediate aggregation in the German cockroach. Proc. Natl. Acad. Sci. U. S. A. 112, 15678–15683.

Wakil, W., Usman Ghazanfar, M., Nasir, F., 2012. Insecticidal efficacy of Azadirachta indica, nucleopolyhedrovirus and chlorantraniliprole singly or combined against field populations of Helicoverpa armigera Hübner (Lepidoptera: Noctuidae). Chil. J. Agr. Res. 72 (1), 53–61.

Wang, S.B., Qu, S., 2017. Insect commensal bacteria and their application prospect in pest control and prevention. CAS 8, 008.

Watanabe, H., Tokuda, G., 2010. Cellulolytic systems in insects. Annu. Rev. Entomol. 55, 609–632.

Wei, G., Lai, Y., Wang, G., Chen, H., Li, F., Wang, S., 2017. Insect pathogenic fungus interacts with the gut microbiota to accelerate mosquito mortality. Proc. Natl. Acad. Sci. U. S. A. 114 (23), 5994–5999.

Werren, J.H., 1997. Biology of *wolbachia*. Annu. Rev. Entomol. 42 (1), 587–609.

Werren, J.H., Baldo, L., Clark, M.E., 2008. *Wolbachia*: master manipulators of invertebrate biology. Nat. Rev. Microbiol. 6 (10), 741–751.

Whitcomb, R.F., 1981. The biology of *spiroplasmas*. Ann. Rev. Entomol. 26 (1), 397–425.

Wolin, M.J., Miller, T.L., 1988. Microbe–microbe interactions. In: Hobson, P.N. (Ed.), The Rumen Microbial Ecosystem. Elsevier Applied Science, London, pp. 343–359.

Wu, X., Appel, A.G., 2017. Insecticide resistance of several field-collected German cockroach (Dictyoptera: Blattellidae) strains. J. Econ. Entomol. 110, 1203–1209.

Xia, X., Zheng, D., Zhong, H., Qin, B., Gurr, G.M., Vasseur, L., et al., 2013. Dna sequencing reveals the midgut microbiota of diamondback moth, *plutella xylostella* (l.) and a possible relationship with insecticide resistance. PLoS One 8 (7), e68852.

Xu, H.X., Zheng, X.S., Liu, S.P., Ye, G.Y., Lv, Z.X., 2009. The role of endosymbionts in insect host resistance against adverse factors. Chin. Bull. Entomol. 46 (3), 350–354.

Yang, H., Peng, J.X., Liu, K.Y., Hong, H.Z., 2006. Diversity and function of symbiotic microbes in the gut of lower termites. Wei Sheng Wu Xue Bao 46 (3), 496–499.

Yang, Y.T., Guo, J.Y., Long, C.Y., Liu, H., Wan, F.H., 2014. Advances in endosymbionts and their functions in insects. Acta Entomol. Sin. 57 (1), 111–122.

Yang, H.T., Zou, S.S., Zhai, L.J., Wang, Y., Zhang, F.M., An, L.G., Yang, G.W., 2017. Pathogen invasion changes the intestinal microbiota composition and induces innate immune responses in the zebrafish intestine. Fish. Shellfish. Immun. 71, 35–42.

Yang, Z., Zhao, J., Wei, N., Feng, M., Xian, M., Shi, X., Zheng, Z., Su, Q., Wong, G.W.K., Li, J., 2018. Cockroach is the major cross-reactive allergen in shrimp-sensitized rural children in South China. Allergy 73 (3), 585–592.

Yukuhiro, F., Miyoshi, T., noda, H., 2014. Actin-mediated transovarial transmission of a yeastlike symbiont in the brown planthopper. J. Insect Physiol. 60 (1), 11–117.

Yun, J.H., Roh, S.W., Whon, T.W., Jung, M.J., Kim, M.S., Park, D.S., 2014. Insect gut bacterial diversity determined by environmental habitat, diet, developmental stage, and phylogeny of host. Appl. Environ. Microbiol. 80 (17), 5254–5264.

Zaidman-Rémy, A., Hervé, M., Poidevin, M., Pili-Floury, S., Kim, M.S., Blanot, D., 2006. The Drosophila, amidase pgrp-lb modulates the immune response to bacterial infection. Immunity 24 (4), 463–473.

Zchori-Fein, E., Gottlieb, Y.S., Kelly, E., Brown, J.K., Wilson, J.M., Karr, T.L., Hunter, M.S., 2001. A newly discovered bacterium associated with parthenogenesis and a change in host selection behavior in parasitoid wasps. Proc. Natl. Acad. Sci. U. S. A. 22, 12555–12683.

Zhang, F., 2012. Identification and anti-fungal activity of intestinal bacterium BGI-17 from *Blattella germanica*. Chin. J. Vector. Biol. Control 23 (1), 39–41.

Zhang, M.M., Chu, Y., Zhao, Z.W., Chun-Ju, A.N., 2012. Progress in the molecular mechanisms of the innate immune responses in insects. Acta Entomol. Sin. 55 (10), 1221–1229.
Zhang, F., Huang, Y.H., Liu, S.Z., Zhang, L., Li, B.T., Zhao, X.X., et al., 2013. Pseudomonas reactans, a bacterial strain isolated from the intestinal flora of *blattella germanica* with anti-beauveria bassiana activity. Environ. Entomol. 42 (3), 453–459.
Zhang, W., Zhang, Y.B., Xue, Y.T., 2017. Research progress of secondary rickettsia in insects. J. Econ. Entomol. 39 (2), 431–443.
Zhang, F., Sun, X.X., Zhanga, X.C., Zhang, S., Lu, L., Xia, Y.M., et al., 2018. The interactions between gut microbiota and entomopathogenic fungi: a potential approach for biological control of *Blattella germanica* (L.). Pest Manag. Sci. 74, 438–447.
Zhao, F., Li, C.B., 2017. The characteristics of enteric bacteria Akkermansia muciniphila and its relationship with the health of the body. Microbiol. Bull. 44 (6), 1458–1463.
Zhao, K., Chang, Z.W., Zang, X.Y., et al., 2012. Research status of symbiotic microbial diversity and its control methods in intestinal termites. Int. J. Dev. Neurosci. 18 (2), 331–337.
Zientz, E., Dandekar, T., Gross, R., 2004. Metabolic interdependence of obligate intracellular bacteria and their insect hosts. Microbiol. Mol. Biol. Rev. 68 (4), 745–770.
Zurek, L., Schal, C., Watson, D.W., 2000. Diversity and contribution of the intestinal bacterial community to the development of Musca domestica (Diptera: Muscidae) larvae. J. Med. Entomol. 37, 924–928.
Zhang, X.Y., Lou, M.F., Shen, W., Fu, R.S., Wang, D.H., 2017. A Maternal low-fiber diet predisposes offspring to improved metabolic phenotypes in adulthood in an herbivorous rodent. Phyisol. Biochem. Zool. 90 (1), 75–84.
Zhang, F.M., Liu, D.Z., Wang, L., Li, T., Chang, Q., An, L.G., Yang, G.W., 2015. Characterization of IgM-binding protein: A pIgR-like molecule expressed by intestinal epithelial cells in the common carp (Cyprinus carpio L.). Vet. Immunol. Immunop 167 (1–2), 30–35.
Zhang, F., Wang, X.J., Huang, Y.H., Zhao, Z.G., Zhang, S.S., Gong, X.S., 2014. Differential expression of hemolymph proteins between susceptible and insecticide-resistant *Blattella germanica* (blattodea: blattellidae). Environ. Entomol. 43 (4), 1117–1123.

FURTHER READING

Atkinson, T.H., Koehler, P.G., Patterson, R.S., 1992. Volatile effects of insect growth regulators against the German cockroach (Dictyoptera: Blattellidae). J. Med. Entomol. 29, 364.
Bosco-Drayon, V., 2013. Peptidoglycan sensing by the receptor PGRP-LE in the drosophila gut induces immune responses to infectious bacteria and tolerance to microbiota. Cell Host Microbe 12 (3), 153–165.
Buchon, N., et al., 2009. Invasive and indigenous microbiota impact intestinal stem cell activity through multiple pathways in Drosophila. Genes Dev. 23, 2333–2344.
Dou, H., Jiang, H., Peng, P., 2013. pH-dependent self-assembly: micellization and micelle–hollow–sphere transition of cellulose-based copolymers. Angew. Chem. 42, 1516–1519.
Etebari, K., Furlong, M.G., Asgari, S., 2015. Genome wide discovery of long intergenic non-coding RNAs in diamondback moth (Plutella xylostella) and their expression in insecticide resistant strains. Sci. Rep. 5, 968–975.
Huang, M., Ma, Y., Wan, Y., 2015. A sensor-software based on a genetic algorithm-based neural fuzzy system for modeling and simulating a wastewater treatment process. Appl. Soft Comput. 27, 1–10.

Kevan, P.G., Kapongo, J.P., Al-Mazra'awi, M., 2018. Honey bees, bumble bees, and biocontrol. In: Bee Pollination in Agricultural Eco-systems. , pp. 65–81.
Liu, Q., Yang, F., 2005. Research on neural network control model in water supply canal system. China Rural Water Hydropower.
Liu, L., Huang, X.L., Qiao, G.X., 2013. Advances in Buchnera aphidicola, a symbiotic symbiont of aphids. Acta Entomol. Sin. 5, 031.
Lopes, R.B., Alves, S.B., 2011. Differential susceptibility of adults and nymphs of *Blattella germanica* (L.) (Blattodea: Blattellidae) to infection by *Metarhizium anisopliae* and assessment of delivery strategies. Neotrop. Entomol. 40, 368–374.
Mendoza, A.G., Al, E., 2012. Absceso hepático amibiano experimental y candidiasis oportunista. In: Sepúlveda, B., Diamond, L.S. (Eds.), Memorias de la Conferencia Internacional sobre Amibiasis (México October 27–29, 1975), *México*, pp. 558–564.
Minard, G., Tran, F.H., Raharimalala, F.N., et al., 2013. Prevalence, genomic and metabolic profiles of Acinetobacter and Asaia associated with field-caught Aedes albopictus from Madagascar. FEMS Microbiol. Ecol. 83 (1), 63–73.
Moran, N.A., Jarvik, T., 2010. Lateral transfer of genes from fungi underlies carotenoid production in aphids. Science 328 (5978), 624–627.
Moran, N.A., Mccutcheon, J.P., Nakabachi, A., 2008. Genomics and evolution of heritable bacterial symbionts. Annu. Rev. Genet. 165–190.
Moya, A., Pereto, J., Gil, R., Latorre, A., 2008. Learning how to live together: genomic insights into prokaryote–animal symbioses. Nat. Rev. Genet. 9, 218–229.
Perlman, S.J., Hunter, M.S., Zchorifein, E., 2006. The emerging diversity of rickettsia. Proc. Biol. Sci. 273 (1598), 2097–2106.
Shi, Z., Yang, Y., Wang, H., 2011. Breastfeeding of newborns by mothers carrying hepatitis B virus: a meta-analysis and systematic review. Arch. Pediatr. Adolesc. Med. 165, 837–846.
Song, M.K., Cheng, S., 2012. Anomalous pseudocapacitive behavior of a nanostructured, mixed-valent manganese oxide film for electrical energy storage. Nano Lett. 12, 3483.
Srini, K., 2008. Cockroaches: ecology, behavior, and natural history. William J. Bell, Louis M. Roth, and Christine A. Nalepa, Integr. Comp. Biol. 48 (4), 541–543.
Takano, Y., Adachi, S., Okuno, M., 2004. The RING finger protein, RNF8, interacts with retinoid X receptor α and enhances its transcription-stimulating activity. J. Biol. Chem. 279, 18926–18934.
Thompson, G.D., Matten, S., Denholm, I., 2008. The Politics of Resistance Management: Working Towards Pesticide Resistance Management Globally. CABI, pp. 146–165.
Toledo, A.V., Alippi, A.M., de Remes Lenicov, A.M.M., 2011. Growth inhibition of *Beauveria bassiana* by bacteria isolated from the cuticular surface of the corn leafhopper, *Dalbulus maidis* and the *planthopper*, Delphacodes kuscheli, two important vectors of maize pathogens. J. Insect Sci. 11 (1), 29.
Van Arnam, E.B., Currie, C.R., Clardy, J., 2018. Defense contracts: molecular protection in insect-microbe symbioses. Chem. Soc. Rev. 47 (5), 1565–1902.

CHAPTER FIVE

Beyond Baculoviruses: Additional Biotechnological Platforms Based on Insect RNA Viruses

Yongchao Zhao*,†, Jingchen Sun†, Vassiliki Labropoulou*, Luc Swevers*

*Institute of Biosciences and Applications, NCSR "Demokritos", Athens, Greece

†College of Animal Science, South China Agricultural University, Guangzhou, People's Republic of China

Contents

Abstract

The baculovirus expression vector system (BEVS) is currently an important and versatile biotechnological platform for protein expression, particularly for large protein complexes, such as viral-like particles (VLPs). However, baculoviruses seem to show limited potential to be developed as vectors for virus-induced gene silencing, a technique that was originally developed to rapidly analyse gene function in plants. On the other hand, reverse genetics systems of insect RNA viruses can be proposed to function as a biotechnological platform for the production of RNA silencing vectors in insects. RNA viruses that can be considered are nonenveloped viruses with positive-sense ssRNA genome (nodavirus, tetravirus, dicistrovirus, macula-like virus) and with segmented dsRNA genome (cytoplasmic polyhedrosis virus, entomobirnavirus). Engineering of the RNA-dependent RNA polymerase and the viral suppressor of RNAi proteins may accomplish the generation of viral vectors that differ in their persistence/pathogenicity

Advances in Insect Physiology, Volume 55
ISSN 0065-2806
https://doi.org/10.1016/bs.aiip.2018.07.002

levels for multiple applications in RNA silencing (from specific gene function analysis to insect pest control). VLPs based on the capsid shell and associated proteins of viruses with segmented dsRNA genome also could provide a safer alternative for dsRNA delivery. In such case, the BEVS and MultiBac technology may provide an ideal system for production.

1. INTRODUCTION

The baculovirus expression vector system (BEVS) was pioneered more than 30 years ago for heterologous protein expression (Summers, 2006) and since then has gradually progressed to a versatile system that has now achieved mainstream status for diverse applications in biotechnology (van Oers et al., 2015). Baculoviruses exist in two forms, of which the occluded derived virus is used as an insecticide (Lacey et al., 2015) and the budded virus produced by infected insect cells is used in the BEVS system. Applications based on the BEVS include now production of individual proteins and multisubunit protein complexes in insect cells, production of mammalian infectious viruses such as recombinant adeno-associated virus, use as gene transduction vehicles in mammalian cells (BacMam technology; Makkonen et al., 2015) and virus display techniques for immunization, imaging and cell targeting (Kost and Kemp, 2016; Paul et al., 2014). Engineering of baculoviruses has become ever more user friendly, for instance through the bacmid technology (Luckow et al., 1993), with several companies offering commercial kits for protein expression based on baculovirus vectors (Liu et al., 2013). The BEVS has been applied to manufacture on industrial scale several biologicals with as outstanding examples a human papillomavirus vaccine against cervical cancer, a seasonal influenza subunit vaccine for adults, several viral subunit vaccines in the veterinary field as well as the first human gene therapy product to gain approval by the European Medicines Agency (Aponte-Ubillus et al., 2018; Kost and Kemp, 2016).

One of the recent developments in the BEVS is the MultiBac system that has revolutionized the field of production of large eukaryotic protein complexes in high quality and quantity for structural and functional studies (Berger et al., 2004; Pelosse et al., 2017; Sari et al., 2016). MultiBac takes advantage of the large capacity of baculoviruses to incorporate multiple inserts as well as the favourable posttranslational modifications and processing that occur in insect cell lines. Baculoviral genomes, conveniently maintained in *Escherichia coli* hosts as "bacmids" to allow efficient recombineering, have undergone considerable modifications to maximize intact protein complex expression, such as deletion of genes that favour cell lysis and apoptosis and the incorporation of elements to facilitate integration of multiple expression cassettes by Tn7 transposition and Cre/loxP recombination system within the bacterial hosts. Furthermore, an array of DNA plasmid modules is available for rapid assembly of multiple expression cassettes as well as their smooth integration in the bacmids (Pelosse et al., 2017; Sari et al., 2016). One of the major applications of this technology is the production of viral-like particles (VLPs) that can be composed of different viral structural proteins. VLPs mimic live viruses but lack genetic content and are therefore safe and not infectious. VLPs are therefore considered promising candidates for vaccination to combat infectious diseases.

The majority of the BEVS/MultiBac systems are based on the *Autographa californica* multiple nuclear polyhedrosis virus (AcMNPV) and the permissive cell lines Sf9 and Hi5 that can be maintained in serum-free media. However, it has also expanded to recombinant protein expression in silkworm larvae using similar bacmid technology based on *Bombyx mori* nuclear polyhedrosis virus (BmNPV). In this case, the process of generation of recombinant viruses and infection of silkworm larvae is streamlined by methods of high efficiency for recombinant virus selection with minimal background and the use of invasive and conditionally lytic *E. coli* strains to deliver directly the baculovirus genome in recipient cells in the absence of DNA isolation and transfection (Sun et al., 2009, 2010; Yao et al., 2010). The BmNPV/silkworm system has several attractive features compared to the AcMNPV/Sf9 system such as lower costs and higher production levels in larvae compared to cell lines and was used for successful production of rotavirus VLPs and human type II collagen in silkworms (Qi et al., 2016; Yao et al., 2012).

The BEVS is by far the best-known biotechnological platform based on insect cells and has up to now eclipsed other biotechnological applications related to insects, including virus-free protein expression systems based on transformed insect cell lines (Yamaji, 2014). However, for other biotechnological applications such as cell-based screening systems (Douris et al., 2006) and RNA interference (RNAi) technology (see further below), baculoviruses may constitute a less suitable approach. In this review, the prospect of using insect RNA virus-based biotechnological platforms is investigated, with emphasis on the delivery of RNAi. The process of RNAi is triggered by (endogenous or exogenous) dsRNA that is processed by Dicer enzyme to small interfering RNAs (siRNAs) which are loaded in an effector complex called RISC (RNA-induced silencing complex) to degrade homologous RNA target sequences (Mello and Conte, 2004). The property of specific gene silencing by dsRNAs makes RNAi an attractive technique to analyse gene function, particularly in nonmodel organisms (Belles, 2010). Moreover, silencing of essential genes by RNAi is considered as a promising method for pest control (Gordon and Waterhouse, 2007; Swevers and Smagghe, 2012). However, because of their relatively high virulence, even in nonnatural hosts, baculoviruses do not function well as silencing vectors in insects since they cause nonspecific effects on physiology and development (Kontogiannatos et al., 2013). Other recombinant viruses, that cause milder symptoms and are more prone to dsRNA production, such as RNA viruses, may be more effective RNAi vectors, and biotechnological platforms for production of recombinant RNA viruses will be presented later.

2. VIRUS-INDUCED GENE SILENCING IN PLANTS AND THE EXPANSION TO VIGS IN INSECTS

Virus-induced gene silencing (VIGS) was developed in plants as an efficient tool to analyse gene function in plants. During RNA virus replication, replication intermediates with dsRNA structure are formed which are recognized by the RNAi machinery to generate a silencing response that is involved in the antiviral response. On the other hand, if a recombinant virus is used that contains an insert of a cellular gene, efficient silencing of the cellular gene will also be accomplished during recombinant virus replication, resulting in phenotypic effects that reveal the function of the gene (Lu et al., 2003). The method of VIGS for analysing gene function is rapid and simple and can therefore be adapted for high-throughput functional genomics approaches. Pioneered as a highly successful approach in *Nicotiana benthamiana*, VIGS has subsequently expanded to many other plants, including model plants (*Arabidopsis*), crops and trees (Lange et al., 2013; Purkayastha and Dasgupta, 2009; Unver and Budak, 2009). To that end, many different VIGS systems were developed based on both RNA and DNA viruses adapted to specific plant species. VIGS in plants has been used to analyse many plant processes such as growth and development, defense response pathway, abiotic stress response and others (Lange et al., 2013; Purkayastha and Dasgupta, 2009; Unver and Budak, 2009).

Because it is based on genetically modified pathogens, VIGS in plants has the disadvantage that effects of virus infection can interfere with the interpretation of the observed phenotypes. Plant viruses also include viral suppressors of RNAi (VSRs) of variable strength that will influence the silencing efficiency (Lim et al., 2010a,b).

Despite the success of VIGS in plants, the technique has not been explored to a great extent in insects. Baculoviruses have been used to express long inverted repeats that fold into dsRNA structures to cause silencing effects (Huang et al., 2007) but, in their wild-type forms, must be considered too virulent to be used as silencing vectors. When Mediterranean corn borer (*Sesamia nonagrioides*) larvae were infected with AcMNPV expressing RNA hairpins, severe mortality was observed that precluded any phenotypic analysis. Use of the less virulent BmNPV for delivery of RNAi did allow the observation of specific phenotypes but only after filtering out nonspecific aberrations caused by the virus infection in a large population of infected animals (Kontogiannatos et al., 2013). It is therefore suggested that "decapitated" mutant baculovirus vectors that are less virulent could be more suitable to act as VIGS vectors. Baculovirus mutants of the master regulator gene *ie-1* (Efrose et al., 2010) or the viral RNA polymerase subunit *lef-8* (Ioannidis et al., 2016) were described but are considered too weak with respect to their replication capacity to act as efficient RNAi transduction vectors. Deletion and modification of other genes may be more successful. Baculoviruses encode factors that promote apoptosis and lysis, and deletion strains were developed of such factors to achieve higher protein expression in the MultiBac system (Berger et al., 2004). Such deletions may also have a positive impact on dsRNA production. Furthermore, a VSR gene was identified in baculoviruses (Mehrabadi et al., 2015) whose strength could be manipulated to maximize silencing efficiency as was realized for plant viruses (Lim et al., 2010a,b). However, optimal VIGS vectors in insects would require a more limited replication capacity as usually observed in baculoviruses to avoid cellular damage. In addition, recognition by the immune response should be avoided while allowing processing by the RNAi machinery for silencing effects (Kolliopoulou et al., 2017). To meet these conditions, recombinant RNA viruses may be a more suitable approach for development of an insect-specific VIGS system than baculoviruses, as discussed later.

3. VIRUS INFECTIONS IN INSECTS AND RNAi

RNAi is considered the most important immune pathway to control virus infections (Bronkhorst and van Rij, 2014; Gammon and Mello, 2015; Vijayendran et al., 2013). In particular, insects seem to have evolved a separate RNAi pathway, characterized by Dicer-2 (Dcr-2) and Argonaute-2

(Ago-2) and the production of siRNAs, for protection against parasitic nucleic acids, such as transposons and DNA and RNA viruses (Dowling et al., 2016; Wynant et al., 2017). In mosquitoes, also parts of the piRNA pathway have evolved to play a role in antiviral immunity in somatic tissues (Miesen et al., 2016), in contrast to other insects such as *Drosophila*, where piRNAs are restricted to the germ line to defend against transposons (Senti and Brennecke, 2010). For DNA viruses, viral miRNAs are produced as an antiviral mechanism and to regulate the infection cycle to optimize virus replication and virion production (Asgari, 2015; Kolliopoulou and Swevers, 2014).

In plants, there are also specializations in the RNAi pathways for antiviral defense that involve specific (and multiple) Dicer enzymes and Argonaute proteins (Ding and Voinnet, 2007). A major difference between plants and insects, however, is the presence of plant host RNA-dependent RNA polymerase (RdRp) enzymes that will amplify the RNAi signals by the production of secondary viral siRNAs (Burgyan and Havelda, 2011; Szittya and Burgyan, 2013). Furthermore, plants generate systemic silencing signals that can easily spread to neighbouring cells as well as to more distantly localized organs (Melnyk et al., 2011). In insects, by contrast, the mechanisms of spreading of the RNAi signal, such as it is observed during feeding of dsRNA from the gut to the internal organs, or during the establishment of a systemic antiviral RNAi response, are not as well understood and may be less robust as in plants (Ivashuta et al., 2015; Karlikow et al., 2014; Tassetto et al., 2017).

Differences between plants and insects regarding VIGS mechanisms can be illustrated by a comparison of the RNAi processes at play during DNA virus infections. During infections of the DNA viruses Invertebrate Iridescent Virus 6 and the baculovirus AcMNPV, viral siRNAs are mainly generated from complementary transcripts that anneal to form dsRNA structures (Bronkhorst et al., 2012; Jayachandran et al., 2012). In plants, viruses of the *Begomovirus* genus of the Geminiviridae family of DNA viruses are used as VIGS vectors (Turnage et al., 2002). However, while complementary transcripts are produced as is the case for DNA viruses in insects, the major pathway to trigger RNAi is believed to involve the RdRp SGS2/SDE1 and its cofactor SGS3 to convert sense single-stranded viral transcripts to dsRNAs (Muangsan et al., 2004). Thus, production of dsRNAs during DNA virus infections in plants is part of an active antiviral response, while in insects their production may be more coincidental. Production of dsRNA templates by plant RdRps also occurs during RNA virus infections for production of secondary viral siRNAs and amplification of the RNAi response (Ding and Voinnet, 2007).

4. PLATFORMS FOR VIGS VECTORS IN INSECTS

4.1 Flock House Virus

Flock house virus (FHV) is a nonenveloped 30 nm icosahedral virus of the family Nodaviridae. Its genome consists of two capped but nonpolyadenylated single-stranded positive-sense RNA segments that are copackaged in single virions (reviews by Odegard et al., 2010; Venter and Schneemann, 2008). Of the bipartite RNA genome, RNA1 (3.1 kb) encodes protein A or the viral RdRp, while RNA2 (1.4 kb) encodes the viral capsid protein precursor. During the course of infection a subgenomic RNA, RNA3, is generated from RNA1 that produces the protein B2 that acts as a VSR (Chao et al., 2005; Qi et al., 2012).

FHV has been studied intensively as a model system for infection of small positive-sense RNA viruses with respect to virus structure and assembly, nonenveloped viral cell entry, RNA genome replication, specific packaging of two genome segments and the antiviral innate immune response in invertebrate hosts (Odegard et al., 2010; Venter and Schneemann, 2008). In *Drosophila* cells, FHV replication occurs in "spherules" or 50 nm membranous vesicles that are formed by invaginations of the outer mitochondrial membranes (Kopek et al., 2007). Within the spherules, replication intermediates with dsRNA structure are shielded from recognition by the RNAi machinery, while additional protection at the site of replication as well as in the cytoplasm is provided by the B2 protein (Aliyari et al., 2008; Short et al., 2016).

FHV virions are initially assembled as noninfectious "provirions" that consist of 180 copies of the viral capsid precursor alpha protein which encapsulate the two RNA genome fragments. To become infectious particles, provirions undergo an autocatalytic maturation step in which the alpha capsid precursor proteins (44 kDa) are cleaved to produce the mature coat proteins beta (39 kDa) and gamma (4.5 kDa) (Krishna et al., 2003). For cell entry, FHV virions are thought to interact with an (unidentified) cellular receptor and be taken up by endocytosis. The gamma peptides remain associated with the virions and have an amphipathic helix structure that will aid in membrane disruption and cytoplasmic penetration during acidification of endosomal vesicles (Odegard et al., 2009). Interestingly, VLPs produced after expression of the alpha protein also undergo autocatalytic maturation in the absence of viral genomes and could rescue "in *trans*" the infectivity of maturation-defective virions (Walukiewicz et al., 2008).

FHV has an extremely wide host range and can replicate in insects, worms, plants, mammalian cells and yeast (Dasgupta et al., 2003;

Lu et al., 2005; Miller et al., 2001). Thus, production systems of FHV vectors could be established in insect cell lines, baculovirus vectors and plants (Krishna et al., 2003; van Cleef et al., 2014; Zhou and Kearney, 2017). FHV RNA1 replicons constitute self-replicating RNAs that are amplified by the RdRp that they encode. Such replicons can be "launched" following the synthesis of RNA1 transcripts with precise 5′- and 3′-ends such that recognition by viral RdRp is maximized. This is conveniently realized by correct positioning of promoters at the 5′-end and the insertion of ribozyme sequences at the 3′-end. Replication of genomic RNA1 is automatically accompanied by replication of subgenomic RNA3 (Miller et al., 2001). For expression of foreign sequences, insertions at the 3′-end of RNA1 (localized in subgenomic RNA3) are used (Fig. 1A), for instance after the stop codon of the ORF of protein A (Price et al., 2000). In such case, a fluorescent recombinant FHV can be produced if GFP is inserted in-frame with the B2 protein and after the critical residues necessary for VSR activity. To produce infectious virions, a second expression plasmid is necessary for the production of viral capsid precursor protein that needs to be constructed using the same precautions as for the replicon (correct positioning of promoter and ribozyme at 5′- and 3′-ends, respectively). After coreplication of recombinant RNA1 and wild-type RNA2, functional recombinant virions can be produced that function as transduction vectors of GFP in *Drosophila* cells and flies. Moreover, insertion of cellular gene sequences after the ORF of GFP resulted in silencing of (essential) cellular genes after infection with recombinant viruses and could accelerate cell death and fly mortality (Taning et al., 2018). These results indicate the potential for FHV to be used as a VIGS vector in insects. However, since FHV is lethal in flies, more engineering is required to attenuate the virulence of FHV (for instance, through modifications in the VSR gene) and to make it more suitable for VIGS. In VIGS, side effects because of damage by viral replication or expression of virulence factors will interfere with the correct interpretation of the phenotypes and optimization of VIGS vectors will need to find the right balance between viral replication and suppression of RNAi (Kolliopoulou et al., 2017). Alternatively, if FHV is to be used in pest control, high VSR activity may be necessary to achieve high levels of FHV accumulation that will allow it to escape control by RNAi (Lan et al., 2016; Zografidis et al., 2015). Insertion of essential cellular gene sequences could then trigger cell death much faster, resulting in a more efficient pest control agent.

Another approach is based on a defective interfering RNA that is derived from RNA2 (Dasgupta et al., 2003). Because defective interfering RNAs are

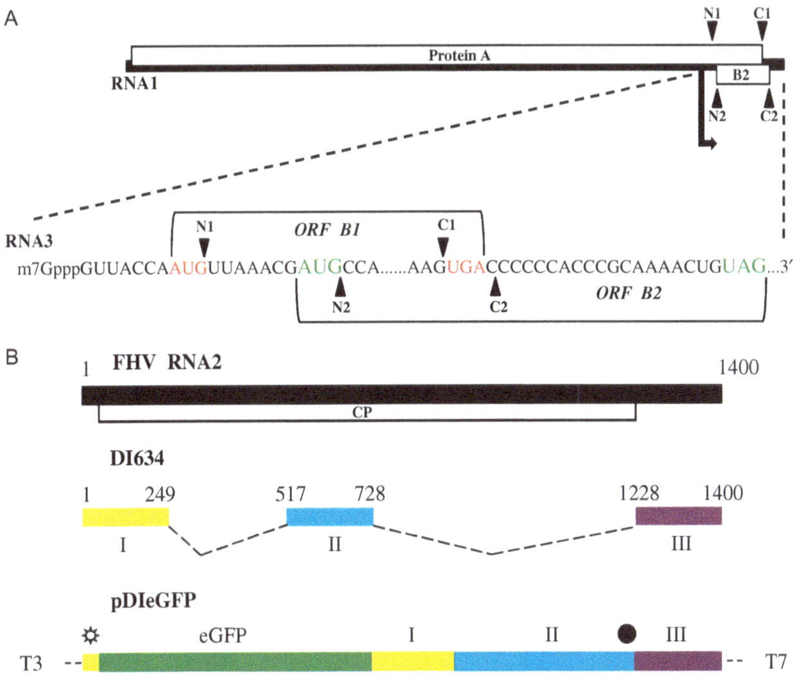

Fig. 1 Strategies for recombinant FHV production. (A) Schematic of RNA1 and subgenomic RNA3. The insertion sites for GFP ORF with respect to the ORF B1 and B2 start codon (N1 and N2) and ORF B1 stop codon (C1 and C2) are indicated. Note that the insertion site C2 is in-frame with the ORF of B2. (B) Schematic of RNA2, its derived defective interfering RNA DI634 and the engineered GFP-expressing defective interfering RNA clone pDIeGFP. The *asterisk* and the *solid circle* show the locations of the start codon and the termination codon for the DI-eGFP fusion, respectively. DI-eGFP is produced after plasmid linearization and T3 RNA polymerase transcription. Abbreviation: *CP*, capsid protein. *Panel (A): Adapted with permission from Price, B.D., Roeder, M., Ahlquist, P., 2000. DNA-directed expression of functional flock house virus RNA1 derivatives in Saccharomyces cerevisiae, heterologous gene expression, and selective effects on subgenomic mRNA synthesis. J. Virol. 74, 11724–11733; panel (B): Adapted, with permission, from Dasgupta, R., Cheng, L.L., Bartholomay, L.C., Christensen, B.M., 2003. Flock house virus replicates and expresses green fluorescent protein in mosquitoes. J. Gen. Virol. 84, 1789–1797.*

recognized by the viral RdRp, they can be amplified to high levels. Insertion of the ORF of GFP at the 5′-end (Fig. 1B) resulted in strong fluorescence after cotransfection with RNA1 replicon. Also functional virions could be produced that were able to transduce fluorescence in mosquito larvae, although this required coexpression of an RNA2 plasmid that encoded functional viral capsid protein (Dasgupta et al., 2003). FHV does not seem to cause adverse effects in mosquitoes and therefore could be applied in more straightforward manner as a transduction or VIGS vector compared to *Drosophila*.

Besides its use as a transduction or VIGS vector, FHV is also used in other nanotechnology applications in the form of VLPs (see also Section 5). High quantities of VLPs based on the viral capsid precursor (alpha) protein can be produced in *Drosophila* cell lines, by baculovirus technology in lepidopteran cells as well as in bacteria (Bajaj and Banerjee, 2016; Krishna et al., 2003; Schneemann and Marshall, 1998). The structure of the capsid protein is characterized by a central β-barrel motif from which several loops are exposed to the surface and that have been used to insert epitopes to develop novel vaccines (Schneemann et al., 2012) or antitoxins (Manayani et al., 2007). Purified VLPs can be loaded with cargo and modified at their surfaces for specific delivery of foreign compounds to particular cell types (Bajaj and Banerjee, 2016). However, an inherent danger in the use of VLPs is the non-specific incorporation of nucleic acids from the cellular system in which they are produced (Krishna et al., 2003). When the RNA content of FHV-derived VLPs, produced by the BEVS, was determined by next-generation sequencing (NGS), mostly ribosomal RNA was found to be incorporated (67%), followed by AcMNPV transcripts (14%) and RNA2 (7.5%) (Routh et al., 2012). However, considerable amounts of transposable elements (5%) and noncoding RNAs (~1%) were also present, raising questions about the safety of using VLP preparations for therapeutic use and advocating the use of NGS as a quality control measure. Interestingly, about 1% of the RNA content of wild-type FHV virions was also identified as host mRNAs, raising the possibility that horizontal gene transfer can occur between hosts of different species after infection with nonenveloped small RNA viruses such as FHV (Routh et al., 2012).

FHV RNA1 genomes engineered with an "origin of assembly" (OA) sequence can also be "trans-encapsidated" by coat proteins of the plant virus tobacco mosaic virus (TMV) (Maharaj et al., 2014). After delivery of TMV particles with trans-encapsidated recombinant FHV RNA1 in mice, an immune response could be detected against GFP, which was expressed from RNA3, indicating successful delivery and expression of recombinant FHV RNA1 in mammalian cells (Zhou et al., 2015). Future improvements in this technology include the induction of RNA replicons that will augment the production of foreign proteins and the efficiency of vaccination (DeFrancesco, 2017).

4.2 Other Single-Stranded Positive-Sense RNA Viruses

4.2.1 *Tetraviridae*

Tetraviruses are currently classified in three families, depending on whether the viral replicase is alphavirus-like (Alphatetraviridae), picornavirus-like (Permutotetraviridae) or carmovirus-like (Carmotetraviridae) (Short et al., 2013). Tetraviruses are characterized by nonenveloped icosahedral virions with a diameter of 35–41 nm and derive their name from the characteristic symmetry $T=4$ of their capsids, which encapsidate either monopartite or bipartite single-stranded positive-sense RNA genomes (Yi et al., 2005).

Tetraviruses are considered to have a very restricted host range that is limited to lepidopteran insects (Hanzlik and Gordon, 1997). Their infectivity is believed to be limited to the larval midgut cells, even when the virus is injected in the haemolymph, and no infection of lepidopteran cell lines was initially reported (Gordon et al., 2001). However, more recently productive infection of Sf9 cells could be achieved after alkaline treatment of virions of *Helicoverpa armigera* stunt virus (HaSV; Alphatetraviridae) (Penkler et al., 2016). These data indicate that the alkaline environment of the lepidopteran insect gut is required for binding and entry of the midgut cells by alphatetraviruses.

On the other hand, Providence virus (PrV), the only member of the Carmotetraviridae isolated so far, was identified as a persistent infection of the *Helicoverpa zea* midgut cell line MG8 (Pringle et al., 2003). PrV seems to be significantly different from other tetraviruses since its replication machinery is derived from plant viruses. PrV may have a wide host range, similar to FHV, since it can infect mammalian cell lines (Jiwaji et al., 2016).

The genome of the type species HaSV and *Nudaurelia* ω virus (NωV; both Alphatetraviridae) constitutes of two RNAs of which RNA1 (~5.5 kb) encodes the replicase protein with similarities to viral RdRps and RNA2 (~2.5 kb) the capsid protein precursor (Fig. 2A; Gordon et al., 2001). A second protein, p17, is also produced from RNA2 that is involved in the selective packaging of viral RNAs (Mendes et al., 2015). Tetravirus infections trigger the RNAi response (*Drosophila* A virus; Wu et al., 2010), but it is not known whether they encode a VSR protein. Similar to FHV, the capsid protein precursor undergoes an autocatalytic maturation step that results in the release of gamma peptides that remain associated with the capsid and are thought to be involved in plasma membrane penetration (although at alkaline pH rather than after acidification; Penkler et al., 2016).

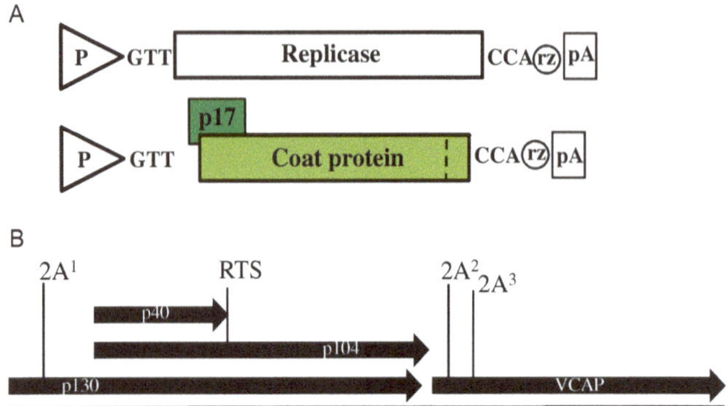

Fig. 2 Tetraviruses. (A) Reverse genetics system for HaSV consists of two plasmids encoding the segment for the replicase (*top*) and the p17/coat protein (*bottom*). The first and final three nucleotides are indicated (GTT and CCA, respectively). Abbreviations: *P*, promoter; *rz*, ribozyme; *pA*, polyadenylation site. (B) Genome organization of PrV. The ORFs of the viral replicase (p40/p104), viral capsid precursor (VCAP) and a non-structural protein of unknown function (p130) as well as 2A-like processing sites are indicated. VCAP is produced from a subgenomic RNA that is generated during viral infection. Abbreviation: *RTS*, read-through stop codon. *Panel (A): Adapted with permission from Gordon, K.H.J., Williams, M.R., Baker, J.S., Gibson, J.M., Bawden, A.L., Millgate, A.G., Larkin, P.J., Hanzlik, T.N., 2001. Replication-independent assembly of an insect virus (Tetraviridae) in plant cells. Virology 288, 36–50; panel (B): Adapted, with permission, from Short, J.R., Nakayinga, R., Hughes, G.E., Walter, C.T., Dorrington, R.A., 2013. Providence virus (family: Carmotetraviridae) replicates vRNA in association with the Golgi apparatus and secretory vesicles. J. Gen. Virol. 94, 1073–1078.*

A reverse genetics system for production of recombinant alphatetraviruses can be realized in similar fashion as for FHV, requiring expression constructs that generate genomic RNA1 and RNA2 with precise terminal ends for efficient packaging (Fig. 2A). While insect cell lines cannot be used because of refractoriness to infection, such system could be realized in plant cells. When the cauliflower mosaic virus 35S promoter was used to express RNA1 and RNA2 of HaSV in *Nicotiana* protoplasts, functional virions could be assembled that were used to infect the midgut of *H. armigera* larvae (Gordon et al., 2001). However, efficient production of HaSV in plants needs optimization; more specifically, suppression of the RNAi response is essential (Gordon and Waterhouse, 2006).

PrV (Carmotetraviridae) has a monopartite genome of 6.5 kb encoding three proteins, including the viral replicase, the capsid protein precursor and a large protein that overlaps with the replicase (Fig. 2B; Jiwaji et al., 2016). The capsid protein precursor is produced from a subgenomic RNA that is generated during infection. A plasmid-based production system is feasible because of demonstrations of viral RNA replication in diverse in vitro transcription–translation systems as well as production of infectious virions in mammalian HeLa cells (Jiwaji et al., 2016). Because PrV infects different insect cell lines, including Sf9 cells, establishment of a reverse system for PrV could also be based on this technology. A study of the kinetics of PrV infection, however, shows that the virus does not establish true persistent infections (as characterized by the absence of cytopathic effects). Instead it appears to kill cells at a low rate, but the pathogenic effect does not spread through the population because of low rate of propagation of infection by an unknown mechanism (Short et al., 2013). While the establishment of a production system for recombinant PrV requires further optimization, the capability of the virus to infect a human cell line may restrict the use of recombinant PrV as a VIGS vector because of safety issues (Jiwaji et al., 2016).

VLPs based on alphatetraviruses can be produced at high level by the BEVS but, similar to FHV, were also found to nonspecifically encapsidate nonviral RNAs (ribosomal RNA, baculovirus transcripts, transposons, noncoding RNAs). Interestingly, different types of transposons are incorporated in NωV VLPs than in FHV VLPs (Routh et al., 2012). Another biotechnological application of tetraviruses is based on the "2A-like" elements that are present in the ORFs and that promote polyprotein processing. The use of 2A peptides is used in multicistronic vectors to achieve expression of equimolar levels of multiple genes from a single construct (e.g. Demidenko et al., 2013).

4.2.2 Dicistroviridae

Dicistroviruses are picorna-like viruses (Picornavirales) characterized by nonenveloped icosahedral virions with a diameter of approximately 30 nm and a pseudo $T=3$ symmetry. Their name is derived from the presence of two ORFs, or cistrons, in the single-stranded positive-sense RNA genome (Bonning and Miller, 2010; Le Gall et al., 2008). The dicistroviral genome has a size of 8500–10,200 nt and contains a 5′-viral protein (VPg) cap and a 3′-poly(A)-tail. Protein synthesis is initiated from two internal ribosomal entry sites (IRES) that are located upstream of ORF1 (5′-untranslated region (UTR) IRES) and ORF2 (intergenic region (IGR) IRES) (Khong et al., 2016). The ORFs encode polyproteins that contain the nonstructural (VSR, helicase, protease, small genome-linked protein (VPg), RdRp, encoded by ORF1) and the structural (capsid-forming) proteins (encoded by ORF2). The polyproteins are cotranslationally and posttranslationally processed by the viral proteases to yield functional units. Viral capsids are assembled by the structural proteins VP0 (=VP4+VP3), VP1 and VP2;

following encapsidation of RNA genome, the capsids undergo a maturation step in which the small VP4 protein is cleaved from VP3 in the VP0 precursor (Mullapudi et al., 2016; Tate et al., 1999).

The type dicistroviruses cricket paralysis virus (CrPV) and *Drosophila* C virus (DCV) have been used as models for picorna-like viruses in basic research with respect to viral capsid structure (Tate et al., 1999), translation initiation by IRES (Lee et al., 2017), the mechanisms of suppression of the formation of stress granules (Khong et al., 2017) and RNAi (Nayak et al., 2010) and the induction of the innate antiviral immune response (Kemp et al., 2013). Other dicistroviruses are studied because of their economic and medical importance: Israeli acute paralysis virus and acute bee paralysis virus are associated with bee decline and colony collapse disorder (Chen et al., 2014; de Miranda et al., 2010); Taura syndrome virus (TSV) is the causative agent of a devastating disease affecting the shrimp farming industry (Mari et al., 2002), while other dicistroviruses are considered as biopesticides against insects of agricultural or medical importance, for instance Triatoma virus (TrV) against triatomine bugs that are vectors of Chagas disease (Bonning and Miller, 2010), and *Homalodisca coagulate virus 1* (HoCV-1) against the glassy-winged sharpshooter, a hemipteran pest (Hunnicutt et al., 2008). Also CrPV is considered as an insect pest control reagent (Manousis and Moore, 1987).

The generation of infectious clones of dicistroviruses is considered an arduous process because of several disadvantageous features (discussed by Carrillo-Tripp et al., 2015). Cloning of the terminal sequences is often difficult because of strong secondary structures leading to artefacts in PCR amplification products. RNA viruses also exist as quasispecies (Lauring and Andino, 2010), and a population consists of different lineages that package genomes of related but not identical sequences. While the population of related genomic RNAs can generate functional viruses efficiently, this is not guaranteed to occur from a single RNA genome that is isolated. In the population, mutants/defects in one RNA lineage are effectively compensated in *trans* by "wild-type" sequences in other lineages of the population. It is therefore recommended to isolate several sequences and test their functionality; in such case RNA genomes that encode proteins with a range of different properties, for instance with respect to replication and RNAi suppression, can also be obtained (Lim et al., 2010a,b). Finally, once isolated, maintenance of complete viral genomes in bacteria can be difficult because of instability and toxic effects (Kerr et al., 2015).

Attempts to isolate infectious clones for black queen cell virus (BQCV) and *Rhopalosiphum padi* virus (RhPV) were finally unsuccessful (Benjeddou et al., 2002; Boyapalle et al., 2014; Pal et al., 2014). Besides the disadvantageous features of RNA viruses in general, generation of reverse genetics systems for dicistroviruses of bees (BQCV) and aphids (RhPV) is more difficult because of the general unavailability of cell lines derived from hymenopteran and hemipteran insects compared to lepidopterans, dipterans and, more recently, coleopterans (Goodman et al., 2012; Lynn, 2001).

More recently, an infectious clone was successfully constructed for CrPV (Kerr et al., 2015). The complete genome was cloned in a plasmid downstream of a T7 RNA polymerase promoter for generation of infectious transcripts that were transfected in the *Drosophila* S2 cell line (Fig. 3A). Infectious virions could be produced that caused mortality in *Drosophila* flies. The infectious clone was used to investigate the effects of modifications in the CrPV genome such as a duplication in the 5′-UTR region and the

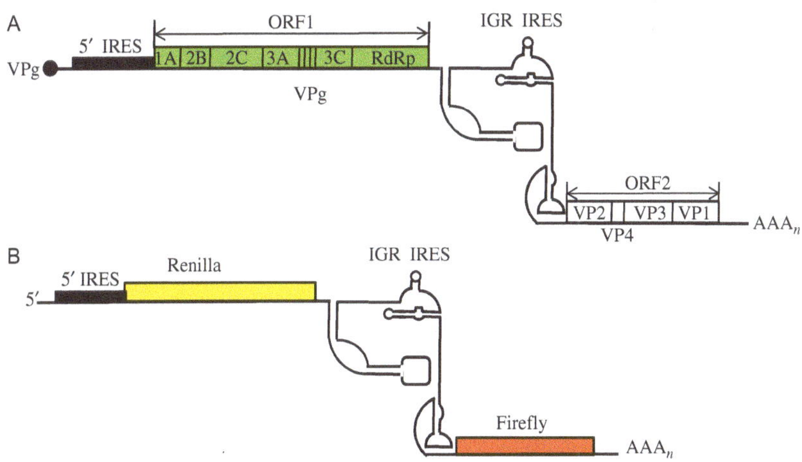

Fig. 3 Schematic of CrPV genome and replicon. (A) CrPV genome organization. Indicated are the viral protein (VPg) cap, 5′- and IGR IRES, ORF1 encoding nonstructural proteins, ORF2 encoding structural (capsid) proteins and poly(A)-tail. (B) In the CrPV double reporter replicon, ORF1 and ORF2 are replaced with Renilla and Firefly ORFs, respectively. Replication and translation is stimulated in *trans* by coexpression of ORF1. *Panel (A): Adapted from Khong, A., Bonderoff, J.M., Spriggs, R.V., Tammpere, E., Kerr, C.H., Jackson, T.J., Willis, A.E., Jan, E., 2016. Temporal regulation of distinct internal ribosome entry sites of the dicistroviridae cricket paralysis virus. Viruses 2016, 8–25 (Open Access); panel (B): Adapted, with permission, from Khong, A., Kerr, C.H., Yeung, C.H.L., Keatings, K., Nayak, A., Allan, D.W., Jan, E., 2017. Disruption of stress granule formation by the multifunctional cricket paralysis virus 1A protein. J. Virol. 91, e01779–16.*

introduction of premature stop codons in the ORFs (Kerr et al., 2015; Khong et al., 2017). The studies also revealed the multifunctional role of the 1A protein which is located at the N-terminus of ORF1. Besides its role as a strong inhibitor of RNAi (Nayak et al., 2010), CrPV-1A has other regulatory functions for establishment of a favourable cellular environment for virus infection such as the inhibition of stress granule formation, which is effected in part through its inhibition of host gene transcription (Khong et al., 2017). For development of CrPV as a platform for production of recombinant protein expression or for VIGS, sites need to be identified that are suitable for insertion of foreign sequences such as reporter genes and RNAi target sequences. Of interest is the observation that viral infections are attenuated by mutations in the 1A protein, providing a platform to test viruses with less pathogenic properties for use as VIGS vectors.

Deletion of ORF2 results in the establishment of CrPV RNA replicons that can be used to drive reporter (luciferase) or recombinant protein expression (Fig. 3B; Khong et al., 2016). Finally, protocols are available for production of (recombinant) CrPV virions at high titers using cell lines such as *Drosophila*-derived DL2 and lepidopteran Hi5 (Scotti et al., 1996). CrPV has a wide host range (infecting insects of six different orders; Bonning and Miller, 2010) and therefore can potentially be applied as a VIGS vector in multiple insect species.

The IGR IRES of Dicistroviridae has caused great interest in basic research of translation initiation since it can recruit the ribosome in the absence of any host initiation factors and initiate translation in the absence of initiator tRNA (Johnson et al., 2017; Lee et al., 2017). The IGR IRES of CrPV, which is studied most intensely, is active in all eukaryotic cells and even in prokaryotes although the mechanism is different (Colussi et al., 2014). The study of IGR IRES of dicistroviruses has contributed immensely to the understanding of the mechanism of translation initiation in eukaryotes. As for practical applications, the IGR IRES can be incorporated in strategies for expression of multiple proteins from a single promoter or transcript (Brodel et al., 2014). Finally, the IRES can be considered as a target for antiviral strategies and the development of antiviral drugs. While such work has already been initiated for picornaviruses that infect humans (Dibrov et al., 2014), it may also be applicable to dicistroviruses that infect insects of economic importance such as honeybees and shrimps.

4.2.3 Macula-Like Latent Virus

Macula-like latent virus (MLV) was identified as a persistent infection in the silkworm (*B. mori*)-derived cell line BmN4 (Katsuma et al., 2005).

Subsequent analysis indicated that MLV is related to plant viruses of the family Tymoviridae (Martelli et al., 2002a) and was acquired by silkworms by an unknown mechanism. Such "jump" from plants to insects is not entirely unexpected since tymoviruses can potentially replicate in both plants and insects (Hogenhout et al., 2008; Martelli et al., 2002b).

The genome of MLV consists of a single-stranded positive-sense RNA of ~6.5 kb (excluding the poly(A)-tail) and contains three ORFs encoding: (1) a large polyprotein (~195 kDa) encoding replicative proteins (methyltransferase, papain-like endopeptidase, helicase, RdRp) characteristic of members of the Tymoviridae; (2) a small capsid protein (25 kDa) and (3) a small protein (15 kDa) with unknown function. The capsid protein and presumably p15 are produced from a subgenomic RNA that is generated during infection (Fig. 4; Katsuma et al., 2005).

During persistent infections of BmN4 cells, especially the subgenomic RNA of MLV can accumulate to high levels and infectious virions of 30 nm are produced. Interestingly, MLV infection activates two antiviral RNAi pathways in BmN4 cells, as documented by the detection of two types of viral small RNAs (Katsuma et al., 2018). SiRNAs with a distinct peak of 20 nt did not display a strong strand bias and could be mapped along the whole length of the genomic RNA. A second class of small RNAs with characteristics of piRNAs (27–28 nt) was also identified with very strong sense bias that almost exclusively mapped to the subgenomic RNA region. Knockdown of different Argonaute proteins confirmed the involvement of both Ago-2 (siRNA pathway) and Argonaute proteins of the Piwi class (Siwi, Ago-3) in antiviral defense in BmN4 cells (Katsuma et al., 2018).

MLV can establish persistent infections in many different lepidopteran cell lines (Swevers et al., 2016) as well as silkworm larvae (Katsuma et al., 2005).

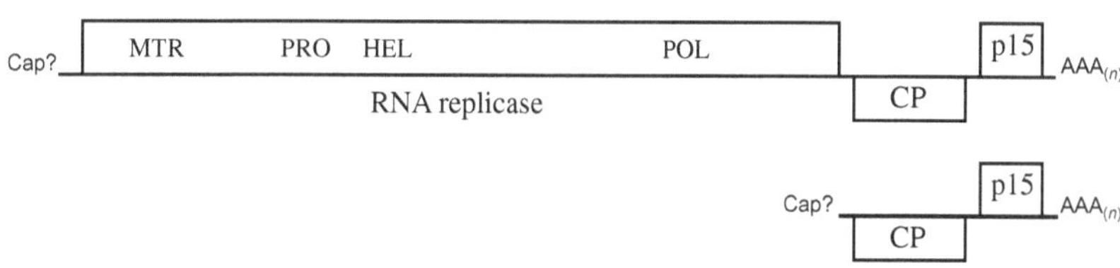

Fig. 4 Genome organization of macula-like virus (MLV). Shown is both the genomic (*top*) and subgenomic RNA (*bottom*). The RNA replicase likely undergoes autocatalytic cleavage to methyltransferase (MTR), papain-like protease (PRO), helicase (HEL) and RdRp (POL) subunits. Both coat protein (CP) and p15 (unknown function) are produced from the subgenomic RNA that is generated during infection. *Adapted, with permission, from Katsuma, S., Tanaka, S., Omuro, N., Takabuchi, L., Daimon, T., Imanishi, S., Yamashita, S., Iwanaga, M., Mita, K., Maeda, S., Kobayashi, M., Shimada, T., 2005. Novel macula-like virus identified in Bombyx mori cultured cells. J. Virol. 79, 5577–5584.*

An infectious clone for MLV was isolated (pHMLV plasmid) that can generate functional virions after transfection of silkworm-derived cell lines (Iwanaga et al., 2012). The infectious clone was used to demonstrate the essential role for the ORF of p15 to establish productive infections (Katsuma et al., 2018). Because the accumulation of high levels of subgenomic RNA, MLV-based vectors have potential for production of recombinant proteins and VIGS. While the host range for MLV needs to be determined, it is noted that similar viruses related to MLV were also isolated from mosquitoes (Wang et al., 2012) and bees (de Miranda et al., 2015). Of interest is also the accumulation of "BeeMLV" in *Varroa* mites which opens the possibility to extend the technology of VIGS to bee parasites.

4.3 Cytoplasmic Polyhedrosis Virus

Cytoplasmic polyhedrosis viruses (CPVs) are common pathogens of insects and have been isolated mainly from Lepidoptera, but also from Diptera, Hymenoptera, Coleoptera, Neuroptera and other arthropods (Belloncik, 1996; Vavra et al., 2016). CPV causes large damage to sericulture in Asia (Jiang and Xia, 2014) but is also being developed as a biological pesticide (Ince et al., 2007; Tan et al., 2008).

CPVs belong to the genus *Cypovirus* of the virus family Reoviridae (subfamily Spinareovirinae) and possess a genome consisting generally of 10 dsRNA segments (Ji et al., 2015). A characteristic feature of CPVs, which is shared with baculoviruses, is the embedment of their virions in crystalline structures called occlusion bodies or polyhedra. Polyhedra of CPVs are mainly formed in the cytoplasm and consist of a single virus-encoded protein, polyhedrin, which is encoded by the smallest segment (segment 10) in their genomes. Based on the electrophoretic pattern of the genome segments, CPVs can be divided in more than 20 different types (or species) which are divergent in the nucleotide and amino-acid sequences of the different segments (Ji et al., 2015). Even for the sequence of the RdRp, which is highest conserved, sequence identities of approximately 30% or less are observed (Graham et al., 2008; Zhou et al., 2014). Furthermore, within a specific type or species, different strains can be distinguished based on the shapes and intracellular localization of the polyhedra (e.g. for *B. mori* CPVs, members of the species type CPV-1; Hagiwara et al., 2002).

CPVs have a wide host range among lepidopterans (Eberle et al., 2012), and CPVs of different evolutionary origin can be isolated from the same host (Zhou et al., 2014). Compared to baculovirus, however, CPVs generally induce chronic and sublethal diseases in the host, which result in reduced fecundity, pupal weight and survival of offspring (Belloncik, 1996). Lepidopteran larvae become infected after oral uptake of polyhedra through feeding. Infection of the midgut epithelium occurs after dissolution of the polyhedra in the alkaline solution of the lepidopteran midgut which frees the occluded virions. Usually infection remains limited to the midgut epithelium although occasionally spread to internal tissues is reported (Kolliopoulou et al., 2015). CPV infections spread in the gut through the release of virions in the gut content, while they spread in the environment through the release of polyhedra with the faeces (Belloncik, 1996).

Cell culture systems for CPV are difficult to establish because of the low efficiency of spread of virus infection from cell to cell (Hill et al., 1999b). Through formation of virion/liposome complexes (virosomes), efficient initial infection of the *Lymantria dispar*-derived Ld652 cell line could be achieved which, however, had only a limited duration. No clear cytopathic effects were observed, and produced polyhedra did not contain functional virions (Hill et al., 1999b). Transfection of CPV segmented dsRNA genomes isolated from purified polyhedra was also insufficient to establish productive CPV infection in silkworm-derived Bm5 cells, although remarkable stability of the transfected dsRNA was observed which could be indicative of low-level genome replication (Swevers et al., 2014). More recently, a reverse genetics system could be established that is based on the BmCPV-SZ strain and silkworm-derived BmN cells (Guo et al., 2018). Pathological infection and polyhedra formation were observed after transfection with transcripts of all 10 segments that were generated in vitro by T7 polymerase. The system was used to create recombinant functional virions expressing the fluorescent protein DsRed instead of polyhedrin and to demonstrate the requirement for segments 2 and 7 for viral replication. The cell culture system can be used to investigate the function of the viral-encoded proteins during infection and for the generation of recombinant viruses with increased insecticidal activity, for instance by the insertion of toxin genes (Guo et al., 2018).

CPV induces a strong RNAi response in silkworm midgut cells that is proportional to the replication level of the viral dsRNA genome segments (Kolliopoulou et al., 2015; Zografidis et al., 2015). The interaction with the RNAi machinery is also obvious from attempts to control CPV infections in transgenic silkworms that express RNA hairpins targeting the CPV genome (Jiang et al., 2017). Besides RNAi, a complex transcriptional response is also observed in the host cells of which the functional (proviral or antiviral) relevance still needs to be investigated in detail (Jiang et al., 2016; Kolliopoulou et al., 2015). Based on a reverse genetics system, recombinant CPV can be employed as a VIGS vector, but the optimal site of insertion of foreign gene sequences still needs to be determined since all genome fragments are essential to establish oral infections (Guo et al., 2018). The host range of the employed CPV type also needs to be established. Interactions between CPV virions and lepidopteran (mainly silkworm) hosts have been investigated intensely during the last years (Gao et al., 2017; Hu et al., 2018; Pan et al., 2017; Wu et al., 2017) and need to be taken into account for evaluation of nonspecific effects in gene silencing experiments. Finally, reoviruses characterized by single-shelled virions that are related to CPVs were also isolated from mosquitoes and a mosquito cell line (*Dinovernavirus* genus; Auguste et al., 2015) and can be developed as VIGS vectors in similar fashion.

The encapsulation into μm-sized occlusion bodies or polyhedra after multimerization of polyhedrin protein is an advantageous feature of CPVs that conveys protection of virions in the environment. Earlier studies established that the incorporation of virions in polyhedra is mediated by amino-terminal sequences of the turret protein (TP; see also Section 5) and the N-terminal α-helix of polyhedrin (Ijiri et al., 2009; Mori et al., 2007). This feature was subsequently used to develop polyhedra-based nanoparticles for the slow release of bioactive materials in applications of stem cell research and regenerative medicine (Nishishita et al., 2011). More recently, however, cryo-electron microscopy (cryo-EM) studies established that the proposed region of TP is actually buried within the protein and that the incorporation of bioactive molecules within polyhedra-based biomaterials was likely a nonspecific process (Chen et al., 2011; Yu et al., 2011). In addition, cryo-EM analysis demonstrated the interaction between polyhedra and virion through the A-spike protein (Chen et al., 2011).

4.4 Arboviruses and Mosquito-Specific RNA Viruses

Arthropod haematophagous vectors of viral disease include mosquitoes, sandflies and ticks that transmit arthropod-borne viruses or arboviruses that belong to different families of RNA viruses such as Flaviviridae, Togaviridae, Bunyaviridae, Reoviridae, and Rhabdoviridae (Weaver and Reisen, 2010). Because of their importance to human health and stock breeding, reverse genetics systems were developed for several arboviruses that support production of recombinant virus in mammalian cell lines. One of the most important systems is based on the Sindbis virus (*Alphavirus*, Togaviridae), a positive-sense, ssRNA virus with genome size of 11.5 kb (Vander Veen et al., 2012). During alphavirus infection, a negative-sense RNA is produced that serves as template for positive-sense genomic RNA as well as a subgenomic 26S RNA from which the structural proteins (viral capsid protein and envelope glycoproteins) are translated (Fig. 5). Alphavirus replicons consist of RNA genomes in which the structural genes are replaced by a foreign gene, achieving high expression in the cytoplasm from the subgenomic promoter. The system also allows the insertion of a second expression cassette

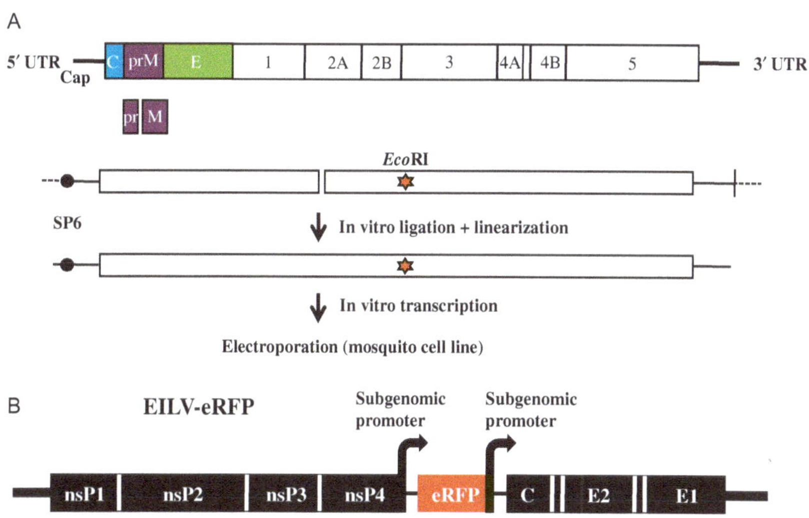

Fig. 5 Mosquito-specific RNA viruses related to arboviruses. (A) Reverse genetics system for Niénokoué virus (NIEV; Flaviviridae). At the *top* the structure of the positive-sense ssRNA genome is shown. The large ORF encodes a polyprotein that is co- and post-translationally cleaved to the three structural proteins (capsid protein, *C*; pre-membrane/membrane protein, *PrM/M*; envelope, *E*) as well as the nonstructural proteins (NS1, NS2A, NS2B, NS3, NS4A, NS4B and NS5). For stable propagation in bacteria, the viral genome is maintained in two plasmids. For production of viral genomic RNA, the two viral cDNA halves are ligated together in vitro prior to linearization and in vitro transcription by SP6 RNA polymerase. Presence/absence of a *Eco*RI site functions as genetic marker. (B) Genomic structure of recombinant Eilat virus (EILV, *Alphavirus*). Nonstructural proteins (nsP1 to nsP4) are produced from the genomic promoter and structural proteins (C, E1, E2) from the subgenomic promoter. In the recombinant EILV, a second subgenomic expression cassette is inserted that drives expression of the fluorescent protein eRFP. *Panel (A): Adapted from Junglen, S., Korries, M., Grasse, W., Wieseler, J., Kopp, A., Hermanns, K., León-Juárez, M., Drosten, C., Kümmerer, B.M., 2017. Host range restriction of insect-specific Flaviviruses occurs at several levels of the viral life cycle. mSphere 2, e00375–16 (open access); panel (B): Adapted, with permission, from Nasar, F., Gorchakov, R.V., Tesh, R.B., Weaver, S.C., 2015. Eilat virus host range restriction is present at multiple levels of the virus life cycle. J. Virol. 89, 1404–1418.*

based on the subgenomic promoter in which case also functional recombinant viral particles are produced capable of foreign gene transduction (Fig. 5B; Lundstrom, 2016). Sindbis-based viral vectors are engineered for safe delivery of RNA replicons expressing viral antigens that are used for vaccine development. Replicon RNA vectors were developed based on alphavirus (e.g. Sindbis virus, Semliki forest virus), flavivirus (e.g. Kunjin virus, dengue virus) and rhabdovirus (e.g. vesicular stomatitis virus) and are produced after T7 RNA polymerase- or RNA polymerase II-mediated transcription (Lundstrom, 2016). Alphavirus replicons can also be delivered in mammalian cells by pseudotyped recombinant baculovirus vectors (Wu et al., 2013).

Recombinant arboviral particles produced in mammalian cells can also infect efficiently mosquito cell lines where they were used as reporter strains and to investigate the interaction with the mosquito host (Attarzadeh-Yazdi et al., 2009; Kohl et al., 1999; Varjak et al., 2018). However, arboviral vectors cannot be easily developed as protein expression in mosquito cell lines or as gene silencing vectors in mosquitoes because of safety issues. On the other hand, mosquito-specific alphaviruses and flaviviruses were also identified that productively infect only mosquitoes and not mammals (Halbach et al., 2017; Roundy et al., 2017). Moreover, several reverse genetics systems based on mosquito-specific RNA viruses that are related to arboviruses were developed recently (Culex flavivirus (CxFV, Flaviviridae), Isawa et al., 2012); Negev virus (NEGV, new *Negevirus* taxon, Gorchakov et al., 2014); Eilat virus (EILV, *Alphavirus*, Nasar et al., 2015); Niénokoué virus (NIEV, Flaviviridae, Junglen et al., 2017; Fig. 5). Because the mosquito-specific viruses do not infect mammalian cells, they provide a more promising platform to be developed as gene transduction or silencing vectors in mosquitoes. Besides the generation of an effective production platform, this work will require also the determination of the host range, the major mode of transmission (horizontal vs vertical) and the virulence of the mosquito-specific RNA viruses. It is noted, however, that, in contrast to the RNA viruses discussed earlier, flaviviruses, alphaviruses, rhabdoviruses and negeviruses are enveloped viruses which make them more complicated vectors and could affect their stability in the environment.

4.5 Plant Viruses Vectored by Insects

Plant viruses vectored by insects present a similar case as arboviruses since they replicate in an organism different from the insect (a plant) and subsequently then use insects (mainly hemipterans but also thysanopterans) for transmission to other plants (Kolliopoulou et al., 2017). However, transmission of plant viruses by insects is believed to occur according to different mechanisms and does not necessarily involve viral replication in the insect vector (as is believed to be the case for arbovirus transmission by mosquitoes; Blanc and Gutierrez, 2015). Modes of transmission by insect vectors include nonpersistent (through interaction with the stylet), semipersistent (through interaction with the foregut) and persistent (transition of the midgut barrier and infection of the salivary gland) (Dietzgen et al., 2016; Whitfield et al., 2015). For persistent (circulative) transmissions, a distinction is made between viruses that also propagate or not propagate in the insect vector. From a biotechnological perspective, persistent circulative infections are the most interesting since invasion of the body cavity is accomplished.

For nonpropagative circulative viruses, viral coat proteins can be engineered to deliver protein cargo into the body cavity after feeding. This was demonstrated for the coat protein of pea enation mosaic virus (PEMV; Luteoviridae), which was expressed as fusion protein with GFP or a peptide toxin in the BEVS or *E. coli*, respectively (Bonning et al., 2014; Whitfield et al., 2014). PEMV coat protein containing a C-terminal extension with the proline-rich region of the read-through domain does not assemble in virions but is expressed as a soluble protein. Despite not forming functional virions, coat protein fusion proteins can efficiently invade the haemocoel of aphids after feeding to cause fluorescence (GFP fusions) or mortality (toxin peptide fusions). Transversion of the midgut barrier was observed for several hemipteran species, including nonhost species, while tobacco budworm larvae (Lepidoptera) were resistant (Bonning et al., 2014). Thus, a biotechnological platform exists for the production of broad spectrum aphicides since the aphid gut barrier is not restrictive for acquisition of luteoviruses (Whitfield et al., 2014). The key for successful production of coat protein fusion proteins is the proline-rich domain that serves as a spacer to allow independent folding of the two protein moieties.

Plant viruses that replicate both in plants and insect vectors include viruses related to Reoviridae, Rhabdoviridae and Bunyaviridae, i.e., families that also harbour arboviruses (Dietzgen et al., 2016; Whitfield et al., 2015). As for arboviruses, plant viruses that are transmitted by insects in a circulative and propagative manner cause disease in the plant but are symptom-free in the insect hosts. Nevertheless, they can trigger a strong antiviral RNAi response in the insect (Lan et al., 2016; Li et al., 2013; Nandety et al., 2013; Xu et al., 2012), which may be involved in the control of the persistent infection as is the case for arboviruses in mosquitoes. A complex reverse genetics system in *Nicotiana benthamiana* plants was recently established for Sonchus yellow net rhabdovirus which was used to investigate the functions of the movement protein and the glycoprotein in the plant, while the capacity to infect aphid vectors was not explored (Wang et al., 2015). Besides plant-based production systems, the availability of (hemipteran or thysanopteran) insect cell lines that support plant virus replication would be very useful. A robust in vitro system for infection of small brown planthopper (*Laodelphax striatellus*) cells by plant viruses of the genus *Fijivirus* (Reoviridae) was recently established, providing a possible platform for the establishment of reverse genetics systems for viruses that cause disease in rice and are vectored by planthoppers (Jia et al., 2012, 2014; Mao et al., 2013). Progress to establish a reverse genetics system for tomato spotted wilt virus (Tospoviridae, Bunyaviriales; vectored by the thrips *Frankliniella occidentalis*) was also reported recently, using yeast as a host (Ishibashi et al., 2017).

5. VLPs AS MORE SAFE ALTERNATIVE FOR DELIVERY OF dsRNA MOLECULES

Heterologous expression of viral capsid proteins in large quantities often results in their spontaneous assembly in "viral-like particles" or VLPs which resemble the outer structure of the virions but lack the incorporation of genetic material. In contrast to functional virions, VLPs do not replicate in host cells or cause disease and therefore can be considered as a more safe strategy to deliver biological materials, such as dsRNA. VLPs may retain the specificity features of natural virions and naturally protect dsRNA from degradation in the environment and the gut. VLPs may therefore be developed as specific nanoparticles to deliver dsRNA to insects (Kolliopoulou et al., 2015).

VLPs that are based on viruses with ssRNA genomes have been demonstrated to package nonspecifically foreign RNA sequences, as documented for FHV and alphatetravirus (Routh et al., 2012). However, they may not package well long dsRNA molecules that have strong secondary structures (Aniagyei et al., 2008). In VLPs based on the bacteriophage Qβ, that naturally packages a positive-sense ssRNA genome within a 30-nm icosahedral capsid, RNAi scaffolds can be incorporated that consist of fusions of a 29-nt Qβ RNA hairpin packaging signal with a 59-nt miRNA-based stem loop (Fang et al., 2017). When the Qβ capsid protein and the RNAi scaffold are coexpressed in bacteria, VLP–RNAi particles are spontaneously assembled and subsequently can be used as delivery vehicles of small RNAs.

After internalization in animal cells by endocytosis, VLPs disassemble and release their cargo. In case of VLP–RNAi particles, Dicer enzyme releases the 20-bp siRNA sequence embedded in the miRNA-like stem loop to trigger specific gene silencing (Fang et al., 2017). The strategy based on short siRNA sequences is favoured for mammalian cells since longer dsRNA sequences are known to nonspecifically activate innate immune response pathways (Elbashir et al., 2001).

In insects, specific gene silencing can be achieved by long dsRNAs for which VLPs based on viruses with dsRNA genomes (such as CPV) may be more suitable. For VLPs based on CPV, research can be guided by the detailed information of CPV virions obtained by cryo-EM (Fig. 6; Hill et al., 1999a; Liu and Cheng, 2015; Yang et al., 2012; Yu et al., 2008, 2011; Zhang et al., 2015; Zhou et al., 2003). Interest in the structure of CPV stems from its

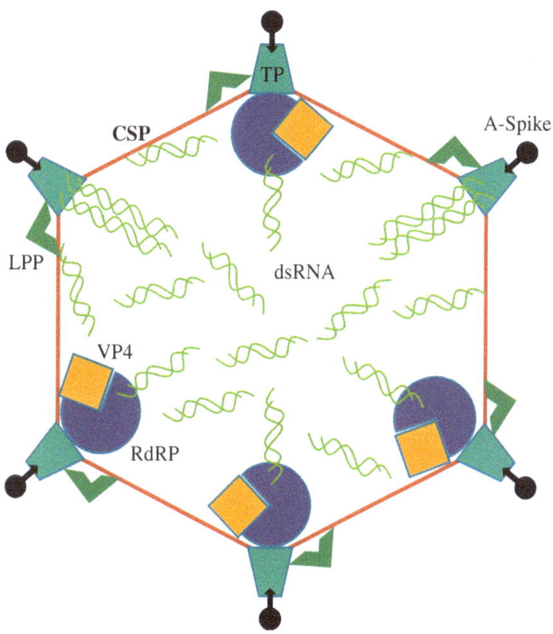

Fig. 6 Schematic of the structure of the CPV virion. The capsid shell is assembled by 120 copies of capsid shell protein (CSP), strengthened by 120 copies of large protruding protein (LPP) and decorated by 12 turrets, each consisting of 5 copies of turret protein (TP; 60 copies per virion). To each turret is attached a copy of A-spike protein (12 copies per virion) at the outside, while inside the shells each of the 10 copies of transcription enzyme complex (TEC) consisting of RdRp and VP4 ATPase is assembled under a turret. Each TEC anchors a dsRNA genomic segment that is organized within the shell in concentric layers.

unique turreted single-layer capsid that resembles the inner core of more complex multilayered reoviruses (Cheng et al., 2011). Furthermore, CPV virions can be isolated in high numbers from polyhedra of infected larvae and be activated in vitro to initiate transcription, thus serving as an easy accessible model for the study of more complex reoviruses of medical and veterinary significance such as rotaviruses and blue-tongue viruses (Yu et al., 2015; Zhang et al., 2015).

Cryo-EM studies established that the CPV virion consists of three major capsid proteins (Fig. 6): (1) 120 molecules of "capsid shell protein" (CSP) (encoded by segment 1) form a thin icosahedral capsid shell of 660 Å in diameter; (2) pentamers of "turret protein" (TP) (encoded by segment 4) form each of the 12 turrets that house the capping complex and (3) 120 "large protrusion proteins" (LPP; encoded by segment 7) interact with each other and CSP to increase capsid stability (Cheng et al., 2011; Yu et al., 2011). The virion also contains 12 copies of the "A-spike" protein (encoded by segment 3), thought to be involved in cell attachment and penetration as well interaction with polyhedrin (Chen et al., 2011; Yu et al., 2011), and 10 copies of the "transcription enzyme complex" (TEC) that consists of the RdRp (segment 2) in complex with the VP4 ATPase (segment 6) (Liu and Cheng, 2015; Zhang et al., 2015). The TECs are located inside the capsid shell at the position of the turrets and anchor the dsRNA genome segments. Despite their simple structure in comparison to other reoviruses, CPV virions are unusually stable and resistant to a variety of chemical treatments such as cations, high pH, proteases, nucleases, phospholipases and SDS (Zhou et al., 2003).

Importantly, expression of high amounts of CSP by the BEVS results in the spontaneous assembly of single-shelled VLPs that are resistant to high pH (Hagiwara and Naitow, 2003). An important question is whether the CSP-based VLPs can be used for efficient delivery of dsRNA or whether additional viral proteins are needed, for instance LPP to increase the stability of the capsid shell. The process of cell entry by CPV is also not well characterized although recent studies have identified candidate proteins that interact with CPV to mediate cell entry, such as integrin β and RACK1 for *B. mori* CPV (Zhang et al., 2017) and alkaline phosphatase for *Dendrolimus punctatus* CPV (Su et al., 2017). In the latter case, involvement of the TP and A-spike proteins was demonstrated, indicating that coexpression of these proteins may be necessary to obtain VLPs with efficient cellular uptake properties. For production of VLPs with different capsid protein composition, the MultiBac system can be employed as demonstrated for other reovirus-based VLPs (Yao et al., 2012).

While turreted CPV virions can have a relatively complicated structure, other viruses with a dsRNA genome, such as birnaviruses (Birnaviridae), resemble more the organization of simple positive-sense ssRNA viruses such as FHV and tetraviruses (Coulibaly et al., 2005). The genome of birnaviruses consists of dsRNA segments of which segment A produces a polyprotein encompassing the capsid protein precursor, a protease and a VSR protein, and segment B the RdRp (Valli et al., 2012; van Cleef et al., 2014). Viruses of the genus *Entomobirnavirus* infect insects (Marklewitz et al., 2012) and may provide an alternative platform for production of VLPs that incorporate dsRNA segments.

6. CONCLUSION

Since its creation, the BEVS has moved from niche status to mainstream technology with respect to protein expression. Especially for large protein complexes, BEVS may become the technology of choice since multiple constructs can be incorporated in high-throughput fashion and strategies can be applied to ensure that the individual expression of subunits is balanced according to the stoichiometry in the complex and specific post-translational modifications are introduced (Sari et al., 2016). The BEVS may therefore have major applications in structural biology and the production of VLPs for vaccine development.

With the discovery of RNAi technology, methods for efficient delivery of siRNA or dsRNA have become ever more important from the viewpoint of potential applications in medicine (e.g. treatment of genetic disease and cancer) and agriculture (e.g. metabolic engineering of crop plants). In insects, RNAi technology has been used to analyse gene function and has been applied as specific insecticide (Kourti et al., 2017; Swevers and Smagghe, 2012). For most insects, however, gene silencing based on application of dsRNA can be very cumbersome and there is a great need to improve the efficiency of delivery. Technology based on RNA viruses may provide a solution since they naturally deliver RNA molecules and are known to trigger the RNAi response. By contrast, baculoviruses may represent a less ideal solution because of their virulence which will cause nonspecific effects in infected (lepidopteran) hosts. Another option may be that the incorporation of RNA hairpin constructs targeting essential genes could increase the virulence of baculovirus vectors for better performance in insect pest control applications. However, also for this application only modest improvements may be expected since RNAi silencing of essential genes may add little additional toxicity during otherwise already virulent infections. Nevertheless, it may be worthwhile to test a range of RNA hairpin constructs whether they could shorten the time to incapacitate the insect pest by recombinant baculoviruses (Kroemer et al., 2015).

In this review, we have explored the capacity of RNA viruses to be developed as RNA silencing vectors in insects. Two potential applications are envisioned. The first would be the use of RNA virus as gene silencing vectors to analyse gene function in insects. While RNAi through dsRNA injection can occur with high efficiency in some insects such as coleopterans and locusts (Wynant et al., 2014), this is not the case for many others such as most lepidopterans (Terenius et al., 2011). Efficient delivery vehicles of RNAi are needed for which RNA viruses can provide a solution. The second application relates to increasing the virulence of RNA viruses in insect pest control. In contrast to baculoviruses and other DNA viruses, it is assumed that RNA viruses trigger a much stronger RNAi response in insects and the incorporation of insect host sequences is therefore expected to have a much greater impact with respect to RNAi effects.

From examining the literature, we propose several reverse genetics systems for RNA viruses that have the potential to be developed as RNA silencing vectors. Proposed RNA viruses include mostly nonenveloped viruses with positive-sense ssRNA genome (FHV, tetraviruses, CrPV, MLV) but also with segmented dsRNA genome (CPV, entomobirnaviruses). In addition, infectious clones for enveloped viruses with positive-sense ssRNA genome that are specific for mosquitoes are available. The further development of these biotechnological platforms will require the assessment of the host range and virulence of these viruses towards different species of insects. Depending on the application (specific gene silencing vs host killing), engineering of the strength of the RdRp and the VSR is considered particularly important since they determine the pathogenic vs persistent level of the infection. Achieving the correct balance between viral replication and suppression of RNAi may be the greatest challenge for the creation of a suitable biotechnological platform for RNA virus-based silencing vectors.

VLPs for delivery of intact dsRNA segments are another potential application. Such VLPs are preferentially based on viruses that naturally package long dsRNA fragments such as CPV. This work can take advantage from the wealth of structural information that has been collected on the CPV virion and that can guide the rational design of VLPs as dsRNA delivery vehicles. Compared to virions, VLPs are considered a safer alternative although their efficiency and specificity need to be assessed. Finally, for the production of VLPs, the BEVS and MultiBac technology may provide an excellent biotechnological platform. While the production of the capsid shell and associated proteins will be relatively straightforward, the loading with dsRNA molecules may be more challenging but could also be accomplished in vitro, after purification of the VLPs.

ACKNOWLEDGEMENTS

L.S. and V.L. acknowledge support by the project "Target Identification and Development of Novel Approaches for Health and Environmental Applications" (MIS 5002514) which is implemented under the Action for the Strategic Development on the Research and Technological Sectors, funded by the Operational Programme "Competitiveness, Entrepreneurship and Innovation" (NSRF 2014-2020) and cofinanced by Greece and the European Union (European Regional Development Fund). Mr. Y.Z. is the recipient of a scholarship under the Graduate Student Overseas Study Program from South China Agricultural University (People's Republic of China, Grant No. 2017LHPY015).

REFERENCES

Aliyari, R., Wu, Q.F., Li, H.W., Wang, X.H., Li, F., Green, L.D., Han, C.S., Li, W.X., Ding, S.W., 2008. Mechanism of induction and suppression of antiviral immunity directed by virus-derived small RNAs in Drosophila. Cell Host Microbe 4, 387–397.

Aniagyei, S.E., DuFort, C., Kao, C.C., Dragnea, B., 2008. Self-assembly approaches to nanomaterial encapsulation in viral protein cages. J. Mater. Chem. 18, 3763–3774.

Aponte-Ubillus, J.J., Barajas, D., Peltier, J., Bardliving, C., Shamlou, P., Gold, D., 2018. Molecular design for recombinant adeno-associated virus (rAAV) vector production. Appl. Microbiol. Biotechnol. 102, 1045–1054.

Asgari, S., 2015. Regulatory role of cellular and viral microRNAs in insect-virus interactions. Curr. Opin. Insect Sci. 8, 104–110.

Attarzadeh-Yazdi, G., Fragkoudis, R., Chi, Y., Siu, R.W.C., Ulper, L., Barry, G., Rodriguez-Andres, J., Nash, A.A., Bouloy, M., Merits, A., Fazakerley, J.K., Kohl, A., 2009. Cell-to-cell spread of the RNA interference response suppresses Semliki Forest virus (SFV) infection of mosquito cell cultures and cannot be antagonized by SFV. J. Virol. 83, 5735–5748.

Auguste, A.J., Kaelber, J.T., Fokam, E.B., Guzman, H., Carrington, C.V.F., Erasmus, J.H., Kamgang, B., Popov, V.L., Jakana, J., Liu, X.A., Wood, T.G., Widen, S.G., Vasilakis, N., Tesh, R.B., Chiu, W., Weaver, S.C., 2015. A newly isolated reovirus has the simplest genomic and structural organization of any reovirus. J. Virol. 89, 676–687.

Bajaj, S., Banerjee, M., 2016. In vitro assembly of polymorphic virus-like particles from the capsid protein of a nodavirus. Virology 496, 106–115.

Belles, X., 2010. Beyond Drosophila: RNAi in vivo and functional genomics in insects. Annu. Rev. Entomol. 55, 111–128.

Belloncik, S., 1996. Interactions of cytoplasmic polyhedrosis viruses with insects. Adv. Insect Physiol. 26 (26), 233–296.

Benjeddou, M., Leat, N., Allsopp, M., Davison, S., 2002. Development of infectious transcripts and genome manipulation of Black queen-cell virus of honey bees. J. Gen. Virol. 83, 3139–3146.

Berger, I., Fitzgerald, D.J., Richmond, T.J., 2004. Baculovirus expression system for heterologous multiprotein complexes. Nat. Biotechnol. 22, 1583–1587.

Blanc, S., Gutierrez, S., 2015. The specifics of vector transmission of arboviruses of vertebrates and plants. Curr. Opin. Virol. 15, 27–33.

Bonning, B.C., Miller, W.A., 2010. Dicistroviruses. Annu. Rev. Entomol. 55, 129–150.

Bonning, B.C., Pal, N., Liu, S.J., Wang, Z.H., Sivakumar, S., Dixon, P.M., King, G.F., Miller, W.A., 2014. Toxin delivery by the coat protein of an aphid-vectored plant virus provides plant resistance to aphids. Nat. Biotechnol. 32, 102–105.

Boyapalle, S., Beckett, R.J., Pal, N., Miller, W.A., Bonning, B.C., 2014. Infectious genomic RNA of Rhopalosiphum padi virus transcribed in vitro from a full-length cDNA clone (375, pg 410, 2008). Virology 452–453, 158.

Brodel, A.K., Sonnabend, A., Kubick, S., 2014. Cell-free protein expression based on extracts from CHO cells. Biotechnol. Bioeng. 111, 25–36.

Bronkhorst, A.W., van Rij, R.P., 2014. The long and short of antiviral defense: small RNA-based immunity in insects. Curr. Opin. Virol. 7, 19–28.

Bronkhorst, A.W., van Cleef, K.W.R., Vodovar, N., Ince, I.A., Blanc, H., Vlak, J.M., Saleh, M.C., van Rij, R.P., 2012. The DNA virus invertebrate iridescent virus 6 is a target of the Drosophila RNAi machinery. Proc. Natl. Acad. Sci. U.S.A. 109, E3604–E3613.

Burgyan, J., Havelda, Z., 2011. Viral suppressors of RNA silencing. Trends Plant Sci. 16, 265–272.

Carrillo-Tripp, J., Bonning, B.C., Miller, W.A., 2015. Challenges associated with research on RNA viruses of insects. Curr. Opin. Insect Sci. 8, 62–68.

Chao, J.A., Lee, J.H., Chapados, B.R., Debler, E.W., Schneemann, A., Williamson, J.R., 2005. Dual modes of RNA-silencing suppression by flock house virus protein B2. Nat. Struct. Mol. Biol. 12, 952–957.

Chen, J., Sun, J.C., Atanasov, I., Ryazantsev, S., Zhou, Z.H., 2011. Electron tomography reveals polyhedrin binding and existence of both empty and full cytoplasmic polyhedrosis virus particles inside infectious polyhedra. J. Virol. 85, 6077–6081.

Chen, Y.P., Pettis, J.S., Corona, M., Chen, W.P., Li, C.J., Spivak, M., Visscher, P.K., DeGrandi-Hoffman, G., Boncristiani, H., Zhao, Y., Vanengelsdorp, D., Delaplane, K., Solter, L., Drummond, F., Kramer, M., Lipkin, W.I., Palacios, G., Hamilton, M.C., Smith, B., Huang, S.K., Zheng, H.Q., Li, J.L., Zhang, X., Zhou, A.F., Wu, L.Y., Zhou, J.Z., Lee, M.L., Teixeira, E.W., Li, Z.G., Evans, J.D., 2014. Israeli acute paralysis virus: epidemiology, pathogenesis and implications for honey bee health. PLoS Pathog. 10, e1004261.

Cheng, L.P., Sun, J.C., Zhang, K., Mou, Z.J., Huang, X.X., Ji, G., Sun, F., Zhang, J.Q., Zhu, P., 2011. Atomic model of a cypovirus built from cryo-EM structure provides insight into the mechanism of mRNA capping. Proc. Natl. Acad. Sci. U.S.A. 108, 1373–1378.

Colussi, T.M., Costantino, D.A., Hammond, J.A., Ruehle, G.M., Nix, J.C., Kieft, J.S., 2014. The structural basis of transfer RNA mimicry and conformational plasticity by a viral RNA. Nature 511, 366–369.

Coulibaly, F., Chevalier, C., Gutsche, I., Pous, J., Navaza, J., Bressanelli, S., Delmas, B., Rey, F.A., 2005. The birnavirus crystal structure reveals structural relationships among icosahedral viruses. Cell 120, 761–772.

Dasgupta, R., Cheng, L.L., Bartholomay, L.C., Christensen, B.M., 2003. Flock house virus replicates and expresses green fluorescent protein in mosquitoes. J. Gen. Virol. 84, 1789–1797.

de Miranda, J.R., Cordoni, G., Budge, G., 2010. The acute bee paralysis virus-Kashmir bee virus-Israeli acute paralysis virus complex. J. Invertebr. Pathol. 103, S30–S47.

de Miranda, J.R., Cornman, R.S., Evans, J.D., Semberg, E., Haddad, N., Neumann, P., Gauthier, L., 2015. Genome characterization, prevalence and distribution of a macula-like virus from Apis mellifera and Varroa destructor. Viruses 7, 3586–3602.

DeFrancesco, L., 2017. The 'anti-hype' vaccine. Nat. Biotechnol. 35, 193–197.

Demidenko, A.A., Blattman, J.N., Blattman, N.N., Greenberg, P.D., Nibert, M.L., 2013. Engineering recombinant reoviruses with tandem repeats and a tetravirus 2A-like element for exogenous polypeptide expression. Proc. Natl. Acad. Sci. U.S.A. 110, E1867–E1876.

Dibrov, S.M., Parsons, J., Carnevali, M., Zhou, S., Rynearson, K.D., Ding, K.J., Sega, E.G., Brunn, N.D., Boerneke, M.A., Castaldi, M.P., Hermann, T., 2014. Hepatitis C virus translation inhibitors targeting the internal ribosomal entry site. J. Med. Chem. 57, 1694–1707.

Dietzgen, R.G., Mann, K.S., Johnson, K.N., 2016. Plant virus-insect vector interactions: current and potential future research directions. Viruses 8, 303.

Ding, S.W., Voinnet, O., 2007. Antiviral immunity directed by small RNAs. Cell 130, 413–426.

Douris, V., Swevers, L., Labropoulou, V., Andronopoulou, E., Georgoussi, Z., Iatrou, K., 2006. Stably transformed insect cell lines: tools for expression of secreted and membrane-anchored proteins and high-throughput screening platforms for drug and insecticide discovery. Adv. Virus Res. 68, 113–156.

Dowling, D., Pauli, T., Donath, A., Meusemann, K., Podsiadlowski, L., Petersen, M., Peters, R.S., Mayer, C., Liu, S.L., Zhou, X., Misof, B., Niehuis, O., 2016. Phylogenetic origin and diversification of RNAi pathway genes in insects. Genome Biol. Evol. 8, 3784–3793.

Eberle, K.E., Jehle, J.A., Huber, J., 2012. Microbial control of crop pests using insect viruses. In: Abrol, D.P., Shankar, U. (Eds.), Integrated Pest Management: Principles and Practice. CABI, pp. 281–298.

Efrose, R., Swevers, L., Iatrou, K., 2010. Baculoviruses deficient in ie1 gene function abrogate viral gene expression in transduced mammalian cells. Virology 406, 293–301.

Elbashir, S.M., Harborth, J., Lendeckel, W., Yalcin, A., Weber, K., Tuschl, T., 2001. Duplexes of 21-nucleotide RNAs mediate RNA interference in cultured mammalian cells. Nature 411, 494–498.

Fang, P.Y., Ramos, L.M.G., Holguin, S.Y., Hsiao, C., Bowman, J.C., Yang, H.W., Williams, L.D., 2017. Functional RNAs: combined assembly and packaging in VLPs. Nucleic Acids Res. 45, 3519–3527.

Gammon, D.B., Mello, C.C., 2015. RNA interference-mediated antiviral defense in insects. Curr. Opin. Insect Sci. 8, 111–120.

Gao, K., Deng, X.Y., Shang, M.K., Qin, G.X., Hou, C.X., Guo, X.J., 2017. iTRAQ-based quantitative proteomic analysis of midgut in silkworm infected with Bombyx mori cytoplasmic polyhedrosis virus. J. Proteomics 152, 300–311.

Goodman, C.L., Stanley, D., Ringbauer, J.A., Beeman, R.W., Silver, K., Park, Y., 2012. A cell line derived from the red flour beetle Tribolium castaneum (Coleoptera: Tenebrionidae). In Vitro Cell. Dev. Biol. Anim. 48, 426–433.

Gorchakov, R.V., Tesh, R.B., Weaver, S.C., Nasar, F., 2014. Generation of an infectious Negev virus cDNA clone. J. Gen. Virol. 95, 2071–2074.

Gordon, K.H.J., Waterhouse, P.M., 2006. Small RNA viruses of insects: expression in plants and RNA silencing. In: Bonning, B.C. (Ed.), Insect Viruses: Biotechnological Applications. Elsevier, San Diego, pp. 459–502.

Gordon, K.H.J., Waterhouse, P.M., 2007. RNAi for insect-proof plants. Nat. Biotechnol. 25, 1231–1232.

Gordon, K.H.J., Williams, M.R., Baker, J.S., Gibson, J.M., Bawden, A.L., Millgate, A.G., Larkin, P.J., Hanzlik, T.N., 2001. Replication-independent assembly of an insect virus (Tetraviridae) in plant cells. Virology 288, 36–50.

Graham, R.I., Morin, B., Lapointe, R., Nealis, V.G., Lucarotti, C.J., 2008. Molecular characterisation of a cypovirus isolated from the western spruce budworm Choristoneura occidentalis. Arch. Virol. 153, 1759–1763.

Guo, R., Cao, G.L., Xue, R.Y., Kumar, D., Chen, F., Liu, W., Jiang, Y., Lu, Y.H., Zhu, L.Y., Liang, Z., Kuang, S.L., Hu, X.L., Gong, C.L., 2018. Exogenous gene can be expressed by a recombinant Bombyx mori cypovirus. Appl. Microbiol. Biotechnol. 102, 1367–1379.

Hagiwara, K., Naitow, H., 2003. Assembly into single-shelled virus-like particles by major capsid protein VP1 encoded by genome segment S1 of Bombyx mori cypovirus 1. J. Gen. Virol. 84, 2439–2441.

Hagiwara, K., Rao, S.J., Scott, S.W., Carner, G.R., 2002. Nucleotide sequences of segments 1, 3 and 4 of the genome of Bombyx mori cypovirus 1 encoding putative capsid proteins VP1, VP3 and VP4, respectively. J. Gen. Virol. 83, 1477–1482.

Halbach, R., Junglen, S., van Rij, R.P., 2017. Mosquito-specific and mosquito-borne viruses: evolution, infection, and host defense. Curr. Opin. Insect Sci. 22, 16–27.

Hanzlik, T.N., Gordon, K.H.J., 1997. The Tetraviridae. Adv. Virus Res. 48, 101–168.

Hill, C.L., Booth, T.F., Prasad, B.V.V., Grimes, J.M., Mertens, P.P.C., Sutton, G.C., Stuart, D.I., 1999a. The structure of a cypovirus and the functional organization of dsRNA viruses. Nat. Struct. Biol. 6, 565–568.

Hill, C.L., Booth, T.F., Stuart, D.I., Mertens, P.P.C., 1999b. Lipofectin increases the specific activity of cypovirus particles for cultured insect cells. J. Virol. Methods 78, 177–189.

Hogenhout, S.A., Ammar, E.D., Whitfield, A.E., Redinbaugh, M.G., 2008. Insect vector interactions with persistently transmitted viruses. Annu. Rev. Phytopathol. 46, 327–359.

Hu, X.L., Zhu, M., Zhang, X., Liu, B., Liang, Z., Huang, L.X., Xu, J., Yu, L., Li, K., Zar, M.S., Xue, R.Y., Cao, G.L., Gong, C.L., 2018. Identification and characterization of circular RNAs in the silkworm midgut following Bombyx mori cytoplasmic polyhedrosis virus infection. RNA Biol. 15, 292–301.

Huang, Y., Deng, F., Hu, Z.H., Vlak, J.M., Wang, H.Z., 2007. Baculovirus-mediated gene silencing in insect cells using intracellularly produced long double-stranded RNA. J. Biotechnol. 128, 226–236.

Hunnicutt, L.E., Mozoruk, J., Hunter, W.B., Crosslin, J.M., Cave, R.D., Powell, C.A., 2008. Prevalence and natural host range of Homalodisca coagulata virus-1 (HoCV-1). Arch. Virol. 153, 61–67.

Ijiri, H., Coulibaly, F., Nishimura, G., Nakai, D., Chiu, E., Takenaka, C., Ikeda, K., Nakazawa, H., Hamada, N., Kotani, E., Metcalf, P., Kawamata, S., Mori, H., 2009. Structure-based targeting of bioactive proteins into cypovirus polyhedra and application to immobilized cytokines for mammalian cell culture. Biomaterials 30, 4297–4308.

Ince, I.A., Demir, I., Demirbag, Z., Nalcacioglu, R., 2007. A cytoplasmic polyhedrosis virus isolated from the pine processionary caterpillar, Thaumetopoea pityocampa. J. Microbiol. Biotechnol. 17, 632–637.

Ioannidis, K., Swevers, L., Iatrou, K., 2016. Bombyx mori nucleopolyhedrovirus lef8 gene: effects of deletion and implications for gene transduction applications. J. Gen. Virol. 97, 786–796.

Isawa, H., Kuwata, R., Tajima, S., Hoshino, K., Sasaki, T., Takasaki, T., Kobayashi, M., Sawabe, K., 2012. Construction of an infectious cDNA clone of Culex flavivirus, an insect-specific flavivirus from Culex mosquitoes. Arch. Virol. 157, 975–979.

Ishibashi, K., Matsumoto-Yokoyama, E., Ishikawa, M., 2017. A tomato spotted wilt virus S RNA-based replicon system in yeast. Sci. Rep. 7, 12647.

Ivashuta, S., Zhang, Y.J., Wiggins, B.E., Ramaseshadri, P., Segers, G.C., Johnson, S., Meyer, S.E., Kerstetter, R.A., McNulty, B.C., Bolognesi, R., Heck, G.R., 2015. Environmental RNAi in herbivorous insects. RNA 21, 840–850.

Iwanaga, M., Hitotsuyama, T., Katsuma, S., Ishihara, G., Daimon, T., Shimada, T., Imanishi, S., Kawasaki, H., 2012. Infection study of Bombyx mori macula-like virus (BmMLV) using a BmMLV-negative cell line and an infectious cDNA clone. J. Virol. Methods 179, 316–324.

Jayachandran, B., Hussain, M., Asgari, S., 2012. RNA interference as a cellular defense mechanism against the DNA virus baculovirus. J. Virol. 86, 13729–13734.
Ji, X.Y., Axford, D., Owen, R., Evans, G., Ginn, H.M., Sutton, G., Stuart, D.I., 2015. Polyhedra structures and the evolution of the insect viruses. J. Struct. Biol. 192, 88–99.
Jia, D.S., Chen, H.Y., Zheng, A.L., Chen, Q., Liu, Q.F., Xie, L.H., Wu, Z.J., Wei, T.Y., 2012. Development of an insect vector cell culture and RNA interference system to investigate the functional role of fijivirus replication protein. J. Virol. 86, 5800–5807.
Jia, D.S., Mao, Q.Z., Chen, H.Y., Wang, A.M., Liu, Y.Y., Wang, H.T., Xie, L.H., Wei, T.Y., 2014. Virus-induced tubule: a vehicle for rapid spread of virions through basal lamina from midgut epithelium in the insect vector. J. Virol. 88, 10488–10500.
Jiang, L., Xia, Q.Y., 2014. The progress and future of enhancing antiviral capacity by transgenic technology in the silkworm Bombyx mori. Insect Biochem. Mol. Biol. 48, 1–7.
Jiang, L., Peng, Z.W., Guo, Y.B., Cheng, T.C., Guo, H.Z., Sun, Q., Huang, C.L., Zhao, P., Xia, Q.Y., 2016. Transcriptome analysis of interactions between silkworm and cytoplasmic polyhedrosis virus. Sci. Rep. 6, 24894.
Jiang, L., Peng, Z.W., Guo, H.Z., Sun, J.C., Sun, Q., Xia, F., Huang, C.L., Xu, G.W., Xia, Q.Y., 2017. Enhancement of antiviral capacity of transgenic silkworms against cytoplasmic polyhedrosis virus via knockdown of multiple viral genes. Dev. Comp. Immunol. 77, 138–140.
Jiwaji, M., Short, J.R., Dorrington, R.A., 2016. Expanding the host range of small insect RNA viruses: Providence virus (Carmotetraviridae) infects and replicates in a human tissue culture cell line. J. Gen. Virol. 97, 2763–2768.
Johnson, A.G., Grosely, R., Petrov, A.N., Puglisi, J.D., 2017. Dynamics of IRES-mediated translation. Philos. Trans. R. Soc. Lond. B Biol. Sci. 372, 20160177.
Junglen, S., Korries, M., Grasse, W., Wieseler, J., Kopp, A., Hermanns, K., León-Juárez, M., Drosten, C., Kümmerer, B.M., 2017. Host range restriction of insect-specific flaviviruses occurs at several levels of the viral life cycle. mSphere 2, e00375-16.
Karlikow, M., Goic, B., Saleh, M.C., 2014. RNAi and antiviral defense in Drosophila: setting up a systemic immune response. Dev. Comp. Immunol. 42, 85–92.
Katsuma, S., Tanaka, S., Omuro, N., Takabuchi, L., Daimon, T., Imanishi, S., Yamashita, S., Iwanaga, M., Mita, K., Maeda, S., Kobayashi, M., Shimada, T., 2005. Novel macula-like virus identified in Bombyx mori cultured cells. J. Virol. 79, 5577–5584.
Katsuma, S., Kawamoto, M., Shoji, K., Aizawa, T., Kiuchi, T., Izumi, N., Ogawa, M., Mashiko, T., Kawasaki, H., Sugano, S., Tomari, Y., Suzuki, Y., Iwanaga, M., 2018. Transcriptome profiling reveals infection strategy of an insect maculavirus. DNA Res. 25, 277–286.
Kemp, C., Mueller, S., Goto, A., Barbier, V., Paro, S., Bonnay, F., Dostert, C., Troxler, L., Hetru, C., Meignin, C., Pfeffer, S., Hoffmann, J.A., Imler, J.L., 2013. Broad RNA interference-mediated antiviral immunity and virus-specific inducible responses in Drosophila. J. Immunol. 190, 650–658.
Kerr, C.H., Wang, Q.S., Keatings, K., Khong, A., Allan, D., Yip, C.K., Foster, L.J., Jan, E., 2015. The 5′ untranslated region of a novel infectious molecular clone of the dicistrovirus cricket paralysis virus modulates infection. J. Virol. 89, 5919–5934.
Khong, A., Bonderoff, J.M., Spriggs, R.V., Tammpere, E., Kerr, C.H., Jackson, T.J., Willis, A.E., Jan, E., 2016. Temporal regulation of distinct internal ribosome entry sites of the dicistroviridae cricket paralysis virus. Viruses 2016, 8–25.
Khong, A., Kerr, C.H., Yeung, C.H.L., Keatings, K., Nayak, A., Allan, D.W., Jan, E., 2017. Disruption of stress granule formation by the multifunctional cricket paralysis virus 1A protein. J. Virol. 91, e01779-16.
Kohl, A., Billecocq, A., Prehaud, C., Yadani, F.Z., Bouloy, M., 1999. Transient gene expression in mammalian and mosquito cells using a recombinant Semliki Forest virus expressing T7 RNA polymerase. Appl. Microbiol. Biotechnol. 53, 51–56.

Kolliopoulou, A., Swevers, L., 2014. Recent progress in RNAi research in Lepidoptera: intracellular machinery, antiviral immune response and prospects for insect pest control. Curr. Opin. Insect Sci. 6, 28–34.

Kolliopoulou, A., Van Nieuwerburgh, F., Stravopodis, D.J., Deforce, D., Swevers, L., Smagghe, G., 2015. Transcriptome analysis of Bombyx mori larval midgut during persistent and pathogenic cytoplasmic polyhedrosis virus infection. PLoS One 10, e0121447.

Kolliopoulou, A., Taning, C.N.T., Smagghe, G., Swevers, L., 2017. Viral delivery of dsRNA for control of insect agricultural pests and vectors of human disease: prospects and challenges. Front. Physiol. 8, 399.

Kontogiannatos, D., Swevers, L., Maenaka, K., Park, E.Y., Iatrou, K., Kourti, A., 2013. Functional characterization of a juvenile hormone esterase related gene in the moth Sesamia nonagrioides through RNA interference. PLoS One 8, e73834.

Kopek, B.G., Perkins, G., Miller, D.J., Ellisman, M.H., Ahlquist, P., 2007. Three-dimensional analysis of a viral RNA replication complex reveals a virus-induced mini-organelle. PLoS Biol. 5, 2022–2034.

Kost, T.A., Kemp, C.W., 2016. Fundamentals of baculovirus expression and applications. In: Vega, M.C. (Ed.), Advanced Technologies for Protein Complex Production and Characterization. Springer International Publishing, pp. 187–197.

Kourti, A., Swevers, L., Kontogiannatos, D., 2017. In search of new methodologies for efficient insect pest control: the RNAi "Movement". In: Shields, V. (Ed.), Entomology. InTech Open, pp. 71–95. ISBN 978-953-51-5041-1.

Krishna, N.K., Marshall, D., Schneemann, A., 2003. Analysis of RNA packaging in wild-type and mosaic protein capsids of flock house virus using recombinant baculovirus vectors. Virology 305, 10–24.

Kroemer, J.A., Bonning, B.C., Harrison, R.L., 2015. Expression, delivery and function of insecticidal proteins expressed by recombinant baculoviruses. Viruses 7, 422–455.

Lacey, L.A., Grzywacz, D., Shapiro-Ilan, D.I., Frutos, R., Brownbridge, M., Goettel, M.S., 2015. Insect pathogens as biological control agents: back to the future. J. Invertebr. Pathol. 132, 1–41.

Lan, H.H., Chen, H.Y., Liu, Y.Y., Jiang, C.Y., Mao, Q.Z., Jia, D.S., Chen, Q., Wei, T.Y., 2016. Small interfering RNA pathway modulates initial viral infection in midgut epithelium of insect after ingestion of virus. J. Virol. 90, 917–929.

Lange, M., Yellina, A.L., Orashakova, S., Becker, A., 2013. Virus-induced gene silencing (VIGS) in plants: an overview of target species and the virus-derived vector systems. In: Becker, A. (Ed.), Virus-Induced Gene Silencing: Methods and Protocols. Humana Press, Totowa, NJ, pp. 1–14.

Lauring, A.S., Andino, R., 2010. Quasispecies theory and the behavior of RNA viruses. PLoS Pathog. 6, e1001005.

Le Gall, O., Christian, P., Fauquet, C.M., King, A.M.Q., Knowles, N.J., Nakashima, N., Stanway, G., Gorbalenya, A.E., 2008. Picornavirales, a proposed order of positive-sense single-stranded RNA viruses with a pseudo-T=3 virion architecture. Arch. Virol. 153, 715–727.

Lee, K.M., Chen, C.J., Shih, S.R., 2017. Regulation mechanisms of viral IRES-driven translation. Trends Microbiol. 25, 547–562.

Li, J.M., Andika, I.B., Shen, J.F., Lv, Y.D., Ji, Y.Q., Sun, L.Y., Chen, J.P., 2013. Characterization of rice black-streaked dwarf virus- and rice stripe virus-derived siRNAs in singly and doubly infected insect vector Laodelphax striatellus. PLoS One 8, e66007.

Lim, H.S., Vaira, A.M., Domier, L.L., Lee, S.C., Kim, H.G., Hammond, J., 2010a. Efficiency of VIGS and gene expression in a novel bipartite potexvirus vector delivery system as a function of strength of TGB1 silencing suppression. Virology 402, 149–163.

Lim, H.S., Vaira, A.M., Reinsel, M.D., Baer, H., Bailey, B.A., Domier, L.L., Hammond, J., 2010b. Pathogenicity of Alternanthera mosaic virus is affected by determinants in RNA-dependent RNA polymerase and by reduced efficacy of silencing suppression in a movement-competent TGB1. J. Gen. Virol. 91, 277–287.
Liu, H.R., Cheng, L.P., 2015. Cryo-EM shows the polymerase structures and a nonspooled genome within a dsRNA virus. Science 349, 1347–1350.
Liu, F.X., Wu, X.D., Li, L., Liu, Z.S., Wang, Z.L., 2013. Use of baculovirus expression system for generation of virus-like particles: successes and challenges. Protein Expr. Purif. 90, 104–116.
Lu, R., Martin-Hernandez, A.M., Peart, J.R., Malcuit, I., Baulcombe, D.C., 2003. Virus-induced gene silencing in plants. Methods 30, 296–303.
Lu, R., Maduro, M., Li, F., Li, H.W., Broitman-Maduro, G., Li, W.X., Ding, S.W., 2005. Animal virus replication and RNAi-mediated antiviral silencing in Caenorhabditis elegans. Nature 436, 1040–1043.
Luckow, V.A., Lee, S.C., Barry, G.F., Olins, P.O., 1993. Efficient generation of infectious recombinant baculoviruses by site-specific transposon-mediated insertion of foreign genes into a baculovirus genome propagated in Escherichia coli. J. Virol. 67, 4566–4579.
Lundstrom, K., 2016. Replicon RNA viral vectors as vaccines. Vaccines (Basel) 4, 39.
Lynn, D.E., 2001. Novel techniques to establish new insect cell lines. In Vitro Cell. Dev. Biol. Anim. 37, 319–321.
Maharaj, P.D., Mallajosyula, J.K., Lee, G., Thi, P., Zhou, Y.Y., Kearney, C.M., McCormick, A.A., 2014. Nanoparticle encapsidation of flock house virus by auto assembly of tobacco mosaic virus coat protein. Int. J. Mol. Sci. 15, 18540–18556.
Makkonen, K.E., Airenne, K., Yla-Herttulala, S., 2015. Baculovirus-mediated gene delivery and RNAi applications. Viruses 7, 2099–2125.
Manayani, D.J., Thomas, D., Dryden, K.A., Reddy, V., Siladi, M.E., Marlett, J.M., Rainey, G.J.A., Pique, M.E., Scobie, H.M., Yeager, M., Young, J.A.T., Manchester, M., Schneemann, A., 2007. A viral nanoparticle with dual function as an anthrax antitoxin and vaccine. PLoS Pathog. 3, 1422–1431.
Manousis, T., Moore, N.F., 1987. Cricket paralysis virus, a potential control agent for the olive fruit-fly, Dacus oleae Gmel. Appl. Environ. Microbiol. 53, 142–148.
Mao, Q.Z., Zheng, S.L., Han, Q.M., Chen, H.Y., Ma, Y.Y., Jia, D.S., Chen, Q., Wei, T.Y., 2013. New model for the genesis and maturation of viroplasms induced by fijiviruses in insect vector cells. J. Virol. 87, 6819–6828.
Mari, J., Poulos, B.T., Lightner, D.V., Bonami, J.R., 2002. Shrimp Taura syndrome virus: genomic characterization and similarity with members of the genus Cricket paralysis-like viruses. J. Gen. Virol. 83, 915–926.
Marklewitz, M., Gloza-Rausch, F., Kurth, A., Kummerer, B.M., Drosten, C., Junglen, S., 2012. First isolation of an entomobirnavirus from free-living insects. J. Gen. Virol. 93, 2431–2435.
Martelli, G.P., Sabanadzovic, S., Abou-Ghanem Sabanadzovic, N., Edwards, M.C., Dreher, T., 2002a. The family Tymoviridae. Arch. Virol. 147, 1837–1846.
Martelli, G.P., Sabanadzovic, S., Ghanem-Sabanadzovic, N.A., Saldarelli, P., 2002b. Maculavirus, a new genus of plant viruses. Arch. Virol. 147, 1847–1853.
Mehrabadi, M., Hussain, M., Matindoost, L., Asgari, S., 2015. The baculovirus antiapoptotic p35 protein functions as an inhibitor of the host RNA interference antiviral response. J. Virol. 89, 8182–8192.
Mello, C.C., Conte, D., 2004. Revealing the world of RNA interference. Nature 431, 338–342.
Melnyk, C.W., Molnar, A., Baulcombe, D.C., 2011. Intercellular and systemic movement of RNA silencing signals. EMBO J. 30, 3553–3563.

Mendes, A., Vlok, M., Short, J.R., Matsui, T., Dorrington, R.A., 2015. An encapsidated viral protein and its role in RNA packaging by a non-enveloped animal RNA virus. Virology 476, 323–333.
Miesen, P., Joosten, J., van Rij, R.P., 2016. PIWIs go viral: arbovirus-derived piRNAs in vector mosquitoes. PLoS Pathog. 12, e1006017.
Miller, D.J., Schwartz, M.D., Ahlquist, P., 2001. Flock house virus RNA replicates on outer mitochondrial membranes in Drosophila cells. J. Virol. 75, 11664–11676.
Mori, H., Shukunami, C., Furuyama, A., Notsu, H., Nishizaki, Y., Hiraki, Y., 2007. Immobilization of bioactive fibroblast growth factor-2 into cubic proteinous microcrystals (Bombyx mori cypovirus polyhedra) that are insoluble in a physiological cellular environment. J. Biol. Chem. 282, 17289–17296.
Muangsan, N., Beclin, C., Vaucheret, H., Robertson, D., 2004. Geminivirus VIGS of endogenous genes requires SGS2/SDE1 and SGS3 and defines a new branch in the genetic pathway for silencing in plants. Plant J. 38, 1004–1014.
Mullapudi, E., Pridal, A., Palkova, L., de Miranda, J.R., Plevka, P., 2016. Virion structure of Israeli acute bee paralysis virus. J. Virol. 90, 8150–8159.
Nandety, R.S., Fofanov, V.Y., Koshinsky, H., Stenge, D.C., Falk, B.W., 2013. Small RNA populations for two unrelated viruses exhibit different biases in strand polarity and proximity to terminal sequences in the insect host Homalodisca vitripennis. Virology 442, 12–19.
Nasar, F., Gorchakov, R.V., Tesh, R.B., Weaver, S.C., 2015. Eilat virus host range restriction is present at multiple levels of the virus life cycle. J. Virol. 89, 1404–1418.
Nayak, A., Berry, B., Tassetto, M., Kunitomi, M., Acevedo, A., Deng, C.H., Krutchinsky, A., Gross, J., Antoniewski, C., Andino, R., 2010. Cricket paralysis virus antagonizes Argonaute 2 to modulate antiviral defense in Drosophila. Nat. Struct. Mol. Biol. 17547-U41.
Nishishita, N., Ijiri, H., Takenaka, C., Kobayashi, K., Goto, K., Kotani, E., Itoh, T., Mori, H., Kawamata, S., 2011. The use of leukemia inhibitory factor immobilized on virus-derived polyhedra to support the proliferation of mouse embryonic and induced pluripotent stem cells. Biomaterials 32, 3555–3563.
Odegard, A.L., Kwan, M.H., Walukiewicz, H.E., Banerjee, M., Schneemann, A., Johnson, J.E., 2009. Low endocytic pH and capsid protein autocleavage are critical components of flock house virus cell entry. J. Virol. 83, 8628–8637.
Odegard, A., Banerjee, M., Johnson, J.E., 2010. Flock house virus: a model system for understanding non-enveloped virus entry and membrane penetration. Curr. Top. Microbiol. Immunol. 343, 1–22.
Pal, N., Boyapalle, S., Beckett, R.J., Miller, W.A., Bonning, B.C., 2014. A baculovirus-expressed dicistrovirus that is infectious to aphids (vol 81, pg 9339, 2007). J. Virol. 88, 3610.
Pan, Z.H., Wu, P., Gao, K., Hou, C.X., Qin, G.X., Geng, T., Guo, X.J., 2017. Identification and characterization of two putative microRNAs encoded by Bombyx mori cypovirus. Virus Res. 233, 86–94.
Paul, A., Hasan, A., Rodes, L., Sangaralingam, M., Prakash, S., 2014. Bioengineered baculoviruses as new class of therapeutics using micro and nanotechnologies: principles, prospects and challenges. Adv. Drug Deliv. Rev. 71, 115–130.
Pelosse, M., Crocker, H., Gorda, B., Lemaire, P., Rauch, J., Berger, I., 2017. MultiBac: from protein complex structures to synthetic viral nanosystems. BMC Biol. 15, 99.
Penkler, D.L., Jiwaji, M., Domitrovic, T., Short, J.R., Johnson, J.E., Dorrington, R.A., 2016. Binding and entry of a non-enveloped T=4 insect RNA virus is triggered by alkaline pH. Virology 498, 277–287.
Price, B.D., Roeder, M., Ahlquist, P., 2000. DNA-directed expression of functional flock house virus RNA1 derivatives in Saccharomyces cerevisiae, heterologous gene expression, and selective effects on subgenomic mRNA synthesis. J. Virol. 74, 11724–11733.

Pringle, F.M., Johnson, K.N., Goodman, C.L., McIntosh, A.H., Ball, L.A., 2003. Providence virus: a new member of the Tetraviridae that infects cultured insect cells. Virology 306, 359–370.

Purkayastha, A., Dasgupta, I., 2009. Virus-induced gene silencing: a versatile tool for discovery of gene functions in plants. Plant Physiol. Biochem. 47, 967–976.

Qi, N., Zhang, L., Qiu, Y., Wang, Z.W., Si, J., Liu, Y.X., Xiang, X., Xie, J.Z., Qin, C.F., Zhou, X., Hu, Y.Y., 2012. Targeting of Dicer-2 and RNA by a viral RNA silencing suppressor in Drosophila cells. J. Virol. 86, 5763–5773.

Qi, Q., Yao, L.G., Liang, Z.S., Yan, D.H., Li, Z., Huang, Y.D., Sun, J.C., 2016. Production of human type II collagen using an efficient baculovirus-silkworm multigene expression system. Mol. Genet. Genomics 291, 2189–2198.

Roundy, C.M., Azar, S.R., Rossi, S.L., Weaver, S.C., Vasilakis, N., 2017. Insect-specific viruses: a historical overview and recent developments. Adv. Virus Res. 98, 119–146.

Routh, A., Domitrovic, T., Johnson, J.E., 2012. Host RNAs, including transposons, are encapsidated by a eukaryotic single-stranded RNA virus. Proc. Natl. Acad. Sci. U.S.A. 109, 1907–1912.

Sari, D., Gupta, K., Balaji, D., Raj, T.G., Aubert, A., Drncova, P., Garzoni, F., Fitzgerald, D., Berger, I., 2016. The MultiBac baculovirus/insect cell expression vector system for producing complex protein biologics. Adv. Exp. Med. Biol. 896, 199–215.

Schneemann, A., Marshall, D., 1998. Specific encapsidation of nodavirus RNAs is mediated through the C terminus of capsid precursor protein alpha. J. Virol. 72, 8738–8746.

Schneemann, A., Speir, J.A., Tan, G.S., Khayat, R., Ekiert, D.C., Matsuoka, Y., Wilson, I.A., 2012. A virus-like particle that elicits cross-reactive antibodies to the conserved stem of influenza virus hemagglutinin. J. Virol. 86, 11686–11697.

Scotti, P.D., Hoefakker, P., Dearing, S., 1996. The production of cricket paralysis virus in suspension cultures of insect cell lines. J. Invertebr. Pathol. 68, 109–112.

Senti, K.A., Brennecke, J., 2010. The piRNA pathway: a fly's perspective on the guardian of the genome. Trends Genet. 26, 499–509.

Short, J.R., Nakayinga, R., Hughes, G.E., Walter, C.T., Dorrington, R.A., 2013. Providence virus (family: Carmotetraviridae) replicates vRNA in association with the Golgi apparatus and secretory vesicles. J. Gen. Virol. 94, 1073–1078.

Short, J.R., Speir, J.A., Gopal, R., Pankratz, L.M., Lanman, J., Schneemann, A., 2016. Role of mitochondrial membrane spherules in flock house virus replication. J. Virol. 90, 3676–3683.

Su, L., Xu, C.R., Cheng, C.G., Lei, C.F., Sun, X.L., 2017. MTase domain of Dendrolimus punctatus cypovirus VP3 mediates virion attachment and interacts with host ALP protein. Viruses 9, 66.

Summers, M.D., 2006. Milestones leading to the genetic engineering of baculoviruses as expression vector systems and viral pesticides. Adv. Virus Res. 68, 3–73.

Sun, J.C., Zhang, E.H., Yao, L.G., Zhang, H.L., Jin, P.F., 2009. A high efficient method of constructing recombinant Bombyx mori (silkworm) multiple nucleopolyhedrovirus based on zero-background TO-mediated transposition in Escherichia coli. Biotechnol. Progr. 25, 524–529.

Sun, J.C., Yao, L.G., Yao, N., Xu, H., Jin, P.F., Kan, Y.C., 2010. Production of recombinant Bombyx mori nucleopolyhedrovirus in silkworm by intrahaemocoelic injection with invasive diaminopimelate auxotrophic Escherichia coli containing BmNPV-Bacmid. Biotechnol. Appl. Biochem. 57, 117–125.

Swevers, L., Smagghe, G., 2012. Use of RNAi for control of insect crop pests. In: Smagghe, G., Diaz, I. (Eds.), Arthropod-Plant Interactions, Novel Insights and Approaches for IPM, Progress in Biological Control. Springer-Verlag, Dordrecht, pp. 177–197.

Swevers, L., Kolliopoulou, A., Li, Z., Daskalaki, M., Verret, F., Kalantidis, K., Smagghe, G., Sun, J.C., 2014. Transfection of BmCPV genomic dsRNA in silkmoth-derived Bm5 cells: stability and interactions with the core RNAi machinery. J. Insect Physiol. 64, 21–29.

Swevers, L., Ioannidis, K., Kolovou, M., Zografidis, A., Labropoulou, V., Santos, D., Wynant, N., Vanden Broeck, J., Wang, L.L., Cappelle, K., Smagghe, G., 2016. Persistent RNA virus infection of lepidopteran cell lines: interactions with the RNAi machinery. J. Insect Physiol. 93–94, 81–93.

Szittya, G., Burgyan, J., 2013. RNA interference-mediated intrinsic antiviral immunity in plants. Curr. Top. Microbiol. Immunol. 371, 153–181.

Tan, L., Zhang, J.M., Li, Y., Li, Y.Q., Jiang, H., Cao, X., Hu, Y.Y., 2008. The complete nucleotide sequence of the type 5 Helicoverpa armigera cytoplasmic polyhedrosis virus genome. Virus Genes 36, 587–593.

Taning, C.N.T., Christiaens, O., Li, X.X., Swevers, L., Casteels, H., Maes, M., Smagghe, G., 2018. Engineered flock house virus for targeted gene suppression through RNAi in fruit flies (*Drosophila melanogaster*) *in vitro* and *in vivo*. Front. Physiol. 9, 805.

Tassetto, M., Kunitomi, M., Andino, R., 2017. Circulating immune cells mediate a systemic RNAi-based adaptive antiviral response in Drosophila. Cell 169, 314–325.

Tate, J., Liljas, L., Scotti, P., Christian, P., Lin, T.W., Johnson, J.E., 1999. The crystal structure of cricket paralysis virus: the first view of a new virus family. Nat. Struct. Biol. 6, 765–774.

Terenius, O., Papanicolaou, A., Garbutt, J.S., Eleftherianos, I., Huvenne, H., Kanginakudru, S., Albrechtsen, M., An, C.J., Aymeric, J.L., Barthel, A., Bebas, P., Bitra, K., Bravo, A., Chevalieri, F., Collinge, D.P., Crava, C.M., de Maagd, R.A., Duvic, B., Erlandson, M., Faye, I., Felfoldi, G., Fujiwara, H., Futahashi, R., Gandhe, A.S., Gatehouse, H.S., Gatehouse, L.N., Giebultowicz, J.M., Gomez, I., Grimmelikhuijzen, C.J.P., Groot, A.T., Hauser, F., Heckel, D.G., Hegedus, D.D., Hrycaj, S., Huang, L.H., Hull, J.J., Iatrou, K., Iga, M., Kanost, M.R., Kotwica, J., Li, C.Y., Li, J.H., Liu, J.S., Lundmark, M., Matsumoto, S., Meyering-Vos, M., Millichap, P.J., Monteiro, A., Mrinal, N., Niimi, T., Nowara, D., Ohnishi, A., Oostra, V., Ozaki, K., Papakonstantinou, M., Popadic, A., Rajam, M.V., Saenko, S., Simpson, R.M., Soberon, M., Strand, M.R., Tomita, S., Toprak, U., Wang, P., Wee, C.W., Whyard, S., Zhang, W.Q., Nagaraju, J., Ffrench-Constant, R.H., Herrero, S., Gordon, K., Swelters, L., Smagghe, G., 2011. RNA interference in Lepidoptera: an overview of successful and unsuccessful studies and implications for experimental design. J. Insect Physiol. 57, 231–245.

Turnage, M.A., Muangsan, N., Peele, C.G., Robertson, D., 2002. Geminivirus-based vectors for gene silencing in Arabidopsis. Plant J. 30, 107–114.

Unver, T., Budak, H., 2009. Virus-induced gene silencing, a post transcriptional gene silencing method. Int. J. Plant Genomics 2009, 198680.

Valli, A., Busnadiego, I., Maliogka, V., Ferrero, D., Caston, J.R., Rodriguez, J.F., Garcia, J.A., 2012. The VP3 factor from viruses of Birnaviridae family suppresses RNA silencing by binding both long and small RNA duplexes. PLoS One 7, e45957.

van Cleef, K.W.R., van Mierlo, J.T., Miesen, P., Overheul, G.J., Fros, J.J., Schuster, S., Marklewitz, M., Pijlman, G.P., Junglen, S., van Rij, R.P., 2014. Mosquito and Drosophila entomobirnaviruses suppress dsRNA- and siRNA-induced RNAi. Nucleic Acids Res. 42, 8732–8744.

van Oers, M.M., Pijlman, G.P., Vlak, J.M., 2015. Thirty years of baculovirus-insect cell protein expression: from dark horse to mainstream technology. J. Gen. Virol. 96, 6–23.

Vander Veen, R.L., Harris, D.L., Kamrud, K.I., 2012. Alphavirus replicon vaccines. Anim. Health Res. Rev. 13, 1–9.

Varjak, M., Dietrich, I., Sreenu, V.B., Till, B.E., Merits, A., Kohl, A., Schnettler, E., 2018. Spindle-E acts antivirally against alphaviruses in mosquito cells. Viruses 10, 88.

Vavra, J., Bily, T., Nebesarova, J., Federici, B.A., 2016. Occurrence, pathology, and ultrastructure of iridovirus and cytoplasmic polyhedrosis viruses in daphnids from the Czech Republic. J. Invertebr. Pathol. 140, 35–38.

Venter, P.A., Schneemann, A., 2008. Recent insights into the biology and biomedical applications of Flock House virus. Cell. Mol. Life Sci. 65, 2675–2687.

Vijayendran, D., Airs, P.M., Dolezal, K., Bonning, B.C., 2013. Arthropod viruses and small RNAs. J. Invertebr. Pathol. 114, 186–195.

Walukiewicz, H.E., Banerjee, M., Schneemann, A., Johnson, J.E., 2008. Rescue of maturation-defective flock house virus infectivity with noninfectious, mature, viruslike particles. J. Virol. 82, 2025–2027.

Wang, L.H., Lv, X.J., Zhai, Y.G., Fu, S.H., Wang, D., Rayner, S., Tang, Q., Liang, G.D., 2012. Genomic characterization of a novel virus of the family Tymoviridae isolated from mosquitoes. PLoS One 7, e39845.

Wang, Q., Ma, X.N., Qian, S.S., Zhou, X., Sun, K., Chen, X.L., Zhou, X.P., Jackson, A.O., Li, Z.H., 2015. Rescue of a plant negative-strand RNA virus from cloned cDNA: insights into enveloped plant virus movement and morphogenesis. PLoS Pathog. 11, e1005223.

Weaver, S.C., Reisen, W.K., 2010. Present and future arboviral threats. Antiviral Res. 85, 328–345.

Whitfield, A.E., Rotenberg, D., German, T.L., 2014. Plant pest destruction goes viral. Nat. Biotechnol. 32, 65–66.

Whitfield, A.E., Falk, B.W., Rotenberg, D., 2015. Insect vector-mediated transmission of plant viruses. Virology 479, 278–289.

Wu, Q.F., Luo, Y.J., Lu, R., Lau, N., Lai, E.C., Li, W.X., Ding, S.W., 2010. Virus discovery by deep sequencing and assembly of virus-derived small silencing RNAs. Proc. Natl. Acad. Sci. U.S.A. 107, 1606–1611.

Wu, Q.F., Xu, F.Q., Fang, L.R., Xu, J.F., Li, B., Jiang, Y.B., Chen, H.C., Xiao, S.B., 2013. Enhanced immunogenicity induced by an alphavirus replicon-based pseudotyped baculovirus vaccine against porcine reproductive and respiratory syndrome virus. J. Virol. Methods 187, 251–258.

Wu, P., Jie, W.C., Shang, Q., Annan, E., Jiang, X.X., Hou, C.X., Chen, T., Guo, X.J., 2017. DNA methylation in silkworm genome may provide insights into epigenetic regulation of response to Bombyx mori cypovirus infection. Sci. Rep. 7, 16013.

Wynant, N., Santos, D., Vanden Broeck, J., 2014. Biological mechanisms determining the success of RNA interference in insects. Int. Rev. Cell Mol. Biol. 312, 139–167.

Wynant, N., Santos, D., Broeck, J.V., 2017. The evolution of animal Argonautes: evidence for the absence of antiviral AGO Argonautes in vertebrates. Sci. Rep. 7, 9230.

Xu, Y., Huang, L.Z., Fu, S., Wu, J.X., Zhou, X.P., 2012. Population diversity of rice stripe virus-derived siRNAs in three different hosts and RNAi-based antiviral immunity in Laodelphgax striatellus. PLoS One 7, e46238.

Yamaji, H., 2014. Suitability and perspectives on using recombinant insect cells for the production of virus-like particles. Appl. Microbiol. Biotechnol. 98, 1963–1970.

Yang, C.W., Jia, G., Liu, H.R., Zhang, K., Liu, G.Q., Sun, F., Zhu, P., Cheng, L.P., 2012. Cryo-EM structure of a transcribing cypovirus. Proc. Natl. Acad. Sci. U.S.A. 109, 6118–6123.

Yao, L.G., Sun, J.C., Xu, H., Kan, Y.C., Zhang, X.M., Yan, H.C., 2010. A novel economic method for high throughput production of recombinant baculovirus by infecting insect cells with Bacmid-containing diminopimelate-auxotrophic Escherichia coli. J. Biotechnol. 145, 23–29.

Yao, L.G., Wang, S.S., Su, S.O., Yao, N., He, J., Peng, L., Sun, J.C., 2012. Construction of a baculovirus-silkworm multigene expression system and its application on producing virus-like particles. PLoS One 7, e32510.
Yi, F.M., Zhang, J.M., Yu, H.Y., Liu, C.F., Wang, J.P., Hu, Y.Y., 2005. Isolation and identification of a new tetravirus from Dendrolimus punctatus larvae collected from Yunnan Province, China. J. Gen. Virol. 86, 789–796.
Yu, X.K., Jin, L., Zhou, Z.H., 2008. 3.88 Angstrom structure of cytoplasmic polyhedrosis virus by cryo-electron microscopy. Nature 453, 415–419.
Yu, X.K., Ge, P., Jiang, J.S., Atanasov, I., Zhou, Z.H., 2011. Atomic model of CPV reveals the mechanism used by this single-shelled virus to economically carry out functions conserved in multishelled reoviruses. Structure 19, 652–661.
Yu, X.K., Jiang, J.S., Sun, J.C., Zhou, Z.H., 2015. A putative ATPase mediates RNA transcription and capping in a dsRNA virus. Elife 4, e07901.
Zhang, X., Ding, K., Yu, X.K., Chang, W., Sun, J.C., Zhou, Z.H., 2015. In situ structures of the segmented genome and RNA polymerase complex inside a dsRNA virus. Nature 527, 531–534.
Zhang, Y.L., Cao, G.L., Zhu, L.Y., Chen, F., Zar, M.S., Wang, S.M., Hu, X.L., Wei, Y.H., Xue, R.Y., Gong, C.L., 2017. Integrin beta and receptor for activated protein kinase C are involved in the cell entry of Bombyx mori cypovirus. Appl. Microbiol. Biotechnol. 101, 3703–3716.
Zhou, Y.Y., Kearney, C.M., 2017. Chimeric flock house virus protein A with endoplasmic reticulum-targeting domain enhances viral replication and virus-like particle trans-encapsidation in plants. Virology 507, 151–160.
Zhou, Z.H., Zhang, H., Jakana, J., Lu, X.Y., Zhang, J.G., 2003. Cytoplasmic polyhedrosis virus structure at 8 angstrom by electron cryomicroscopy: structural basis of capsid stability and mRNA processing regulation. Structure 11, 651–663.
Zhou, Y., Qin, T.C., Xiao, Y.Z., Qin, F.J., Lei, C.F., Sun, X.L., 2014. Genomic and biological characterization of a new Cypovirus isolated from Dendrolimus punctatus. PLoS One 9, e113201.
Zhou, Y.Y., Maharaj, P.D., Mallajosyula, J.K., McCormick, A.A., Kearney, C.M., 2015. In planta production of flock house virus transencapsidated RNA and its potential use as a vaccine. Mol. Biotechnol. 57, 325–336.
Zografidis, A., Van Nieuwerburgh, F., Kolliopoulou, A., Apostolou-Karampelis, K., Head, S.R., Deforce, D., Smagghe, G., Swevers, L., 2015. Viral small-RNA analysis of Bombyx mori larval midgut during persistent and pathogenic cytoplasmic polyhedrosis virus infection. J. Virol. 89, 11473–11486.

CHAPTER SIX

Emerging RNA Suppression Technologies to Protect Citrus Trees From Citrus Greening Disease Bacteria

Wayne B. Hunter*,†, Xiomara H. Sinisterra-Hunter*,†
*USDA-ARS, Fort Pierce, FL, United States
†AgTec, LLC, Port Saint Lucie, FL, United States

Contents

Advances in Insect Physiology, Volume 55
ISSN 0065-2806
https://doi.org/10.1016/bs.aiip.2018.08.001

Abstract

Research on the development of RNA-suppressing technologies, which include RNA interference, (RNAi), and the application of RNA suppression using cell-penetrating peptide Morpholinos, PPMOs, for suppressing pests and pathogens are briefly presented. Due to the extensive nature of RNAi and antisense oligonucleotide research, readers are directed to this review to obtain a more in-depth understanding of each field of study. Genomic resources and websites for dataset downloads and free access are listed to facilitate a broader community collaboration to develop solutions for this complex Liberibacter pathosystem. Strategies for gene drive and psyllid population modification or replacement, with *Wolbachia* or CRISPR/Cas9-based alterations, have continued to gain interest and are being developed (Champer et al., 2016; Gantz and Akbari, 2018; Peters et al., 2016).

ABBREVIATIONS

ACP	Asian citrus psyllid
***ak*-dsRNA**	arginine kinase double-stranded ribonucleic acid
BAPC	branched amphiphilic peptide capsules
C.	*Candidatus*
***C.*P. armatura**	*Candidatus* Profftella armatura
Cas	CRISPR-associated proteins
CLas	*Candidatus* Liberibacter asiaticus
CPP	cell-penetrating peptide
CRISPR	clustered regularly interspaced short palindromic repeats
CRISPRi	CRISPR interference
CSBV	Chinese sacbrood virus
D. citri	*Diaphorina citri*
DIACI	*Diaphorina citri* genome
DNA	deoxyribonucleic acid
DRBD	dsRNA binding domain
dsRNA	double-stranded RNA
E. coli	*Escherichia coli*
Laf	Liberibacter africanus
Lam	Liberibacter americanus
Las	Liberibacter asiaticus
LC50	lethal concentration 50%
mL	millilitres
MO	Morpholino
nmol	nanomoles
nt	nucleotide
PNA	polynucleic acid
PPMO	penetrating peptide Morpholino
PTD	peptide transduction domain
PTD-DRBD	peptide transduction domain-dsRNA binding domain
RNA	ribonucleic acid
RNAi	RNA interference

RNP	ribonucleoprotein particle
siRNA	short interfering RNA
ssRNA	single-stranded RNA
TRX	thioredoxin
UCSF	University of California, San Francisco
ULS	universal linkage system
UPR	unfolded protein response
wDi	*Wolbachia-Diaphorina citri*
ZNF	zinc finger nucleases
μg	microgram

1. ASIAN CITRUS PSYLLID AND CITRUS GREENING DISEASE, HUANGLONGBING

Huanglongbing (HLB), or citrus greening disease, is currently the most destructive disease in citrus growing regions of Asia, Africa, and the Americas (see reviews: da Graça et al., 2016; Wang et al., 2017a). In the USA it has spread throughout the southern states from Florida to California and threatens the sustainability of the US citrus industry (Gottwald, 2010; Halbert and Manjunath, 2006; Hall et al., 2013). Infected citrus trees may not show symptoms for 2 years. Symptoms start with leaves on a branch turning yellow, then the fruit becomes small, lopsided, and bitter. Eventually the tree declines and dies (Bové, 2006). Symptoms caused by Liberibacter bacteria are the result of self-destructive plant responses and/or direct disruption of the phloem function affecting the phloem loading, unloading, and general nutrient transport (see *plant response model*, Wang et al., 2017b). New Liberibacter-associated diseases in a wide variety of crops are being identified and reported. A review of the differences in these Liberibacters and their virulence traits, including secretion systems, putative effectors, and lipopolysaccharides, and other traits which contribute to pathogenicity and epidemiology, e.g., genes, proteins, enzymes, flagella, and prophages to name a few, is thoroughly discussed in Wang et al. (2017a). The Asian citrus psyllid (ACP), *Diaphorina citri* Kuwayama (Hemiptera: Liviidae), is a phloem-feeding insect and the most economically important pest of citrus because of its status as the main psyllid vector of these bacterial agents (Ammar et al., 2018; Hall et al., 2013). The bacteria associated with HLB include three species of bacteria in the *Candidatus* Liberibacter genus: *C.* Liberibacter asiaticus, Las, *C.* Liberibacter americanus, Lam, and C. Liberibacter africanus, Laf (Bové, 2006; Haapalainen, 2014). Four subspecies have also been reported

(Roberts et al., 2015). The bacteria replicate in the host plant and in the psyllid vectors, with infection occurring in the psyllid's salivary glands and midgut tissues (Ammar et al., 2011a,b, 2016; Ghanim et al., 2016; Kruse et al., 2017; Xu et al., 1988). Thus far, in all cases to date one or more psyllid species are associated with transmission of one or more Liberibacter species (Lallemand et al., 1986; Massonié et al., 1976). Thus once a psyllid acquires the pathogen, as either a nymph or adult, it remains infective, often for the life of the insect (Capoor et al., 1974; Grafton-Cardwell et al., 2013; Hall et al., 2013). Both adult and nymph instars can transmit the bacteria once acquired (Pelz-Stelinski and Killiny, 2016; Pelz-Stelinski et al., 2010; Ren et al., 2016). This supports a replicative, circulative transmission modality of the pathogen (Inoue et al., 2009). Low levels of transovarial transmission of Liberibacters have been reported for other psyllid transmitted Liberibacters, like *Trioza erytreae*, the African psyllid, and vector of Liberibacter africanus, Laf (Aubert, 1987; van den Berg et al., 1992); and potato psyllid, *Bactericera cockerelli*, the vector of L. solanacearum (a.k.a. L. psyllaurous) (Hansen et al., 2008; Munyaneza et al., 2012). Since a female may produce up to 750 eggs per season, the numbers of infected psyllids can increase very rapidly (Tsai and Liu, 2000).

Current management strategies to reduce the psyllid populations have only been moderately successful, depending heavily upon the use of insecticides (Tiwari et al., 2011; Yamamoto et al., 2009); Eulophid parasitoids, *Tamarixia radiata* (Hymenoptera: Eulophidae) (Hoddle and Pandey, 2014); predators like ladybird beetles (Coleoptera) (Michaud and Olsen, 2004); and Entomopathogenic fungi (Avery et al., 2009, 2011; Hunter et al., 2011; Patt et al., 2015). Hunter et al. (2009) provided evidence for multiple species of viruses, including an unknown type of ssRNA, deformed wing virus, and validated a Reovirus species (Marutani-Hert et al., 2008, 2009) which caused infection but only slight negative effects on *D. citri* in Florida, USA. Following these reports, Nouri et al. (2015) confirmed and expanded the number of virus sequences isolated from psyllid collections from Florida, California, and Hawaii in USA, along with samples from China, Taiwan, Brazil, and Pakistan. These new viruses have significant sequence similarity to several distinct taxa including the Reoviridae, Bunyaviridae, Parvoviridae, multiple families within the Picornavirus superfamily, and to many unclassified positive-sense single-stranded RNA viruses. The uses of viruses as biological control agents, cellular cargo transporters, or expression vectors for suppression products, and RNAi, are well known to molecular insect virologists (Kolliopoulou et al., 2017; Lacey et al., 2015). While traditional breeding of resistant varieties is the most sustainable approach for controlling most plant diseases, HLB-resistant citrus variety development has been hampered by lack of HLB-resistant germplasm, polyembryony, pollen–ovule sterility, sexual and graft incompatibilities, and extended juvenility (Davey et al., 2005). Among the transgenic approaches that have been tried to improve citrus resistance against HLB include: (1) overexpression antimicrobial proteins (e.g. thionins and plant defensins) (Hao et al., 2016); (2) overexpressing NPR1 (nonexpressor of pathogenesis-related genes) (Dutt et al., 2015); and (3) antimicrobial peptides (e.g. cecropin B) (Stover et al., 2013). Citrus varieties developed by Dr Erik Mirkov (Texas A&M University) expressing spinach defensins are in field trials under a permit from the US Environmental Protection Agency, EPA (Stover et al., 2013). The regulatory pathway to register transgenic citrus is well established and takes time, averaging 10–12 years. The primary economic concern is consumer acceptance, which requires investments in early education, as well as adult and public education on the facts surrounding important issues of plant improvement. The facts on safety, nutrition, and economic benefits are well documented with over 30 years for GMO crops like maize, soybean, and others (Baum and Roberts, 2014; Bawa and Anilakumar, 2013; Brookes and Barfoot, 2016; Ivashuta et al., 2009, 2015; James, 2015; Koch et al., 2015; Kyndt et al., 2015; Petrick et al., 2013, 2016; Ricroch and Hénard-Damave, 2016; Saurabh et al., 2014; Sherman et al., 2015; Wise, 2013; Wise et al., 2013; Younis et al., 2014). So what other strategies can be applied to suppress the psyllid vector and the bacterial pathogen(s) which affect citrus sustainability?

1.1 Power of RNA Suppression

RNA interference (RNAi) has provided a new understanding of the world around us (Kupferschmidt, 2013; Kyndt et al., 2015; Whyard et al., 2015). RNAi methodologies are used to conduct functional genomics studies in traditionally difficult to work with organisms allowing for the transference of promising results from model systems into the realm of practical applications (Brookes and Barfoot, 2016; Petrick et al., 2016; Ricroch and Hénard-Damave, 2016; Sinisterra-Hunter and Hunter, 2018). Early reviews touted RNAi as the new biopesticide, safer, biodegradable, with increased pest specificity, which uses naturally occurring molecules—dsRNA, siRNA (Bellés, 2010; Huvenne and Smagghe, 2010; Jinak and Doudna, 2009; Whyard et al., 2009). As the understanding of cellular functional genomics advances, the insights into form and function of RNAs are influencing the vision of how to use those molecules to solve complex biological problems.

Being able to alter or modulate gene expression in a manner which mimics the natural molecular cell processes enables development of treatments to reduce pests and pathogen while improving the health of humans, animals, and plants (Marshall and Akbari, 2016, 2018; Sinisterra-Hunter and Hunter, 2018) (Fig. 1).

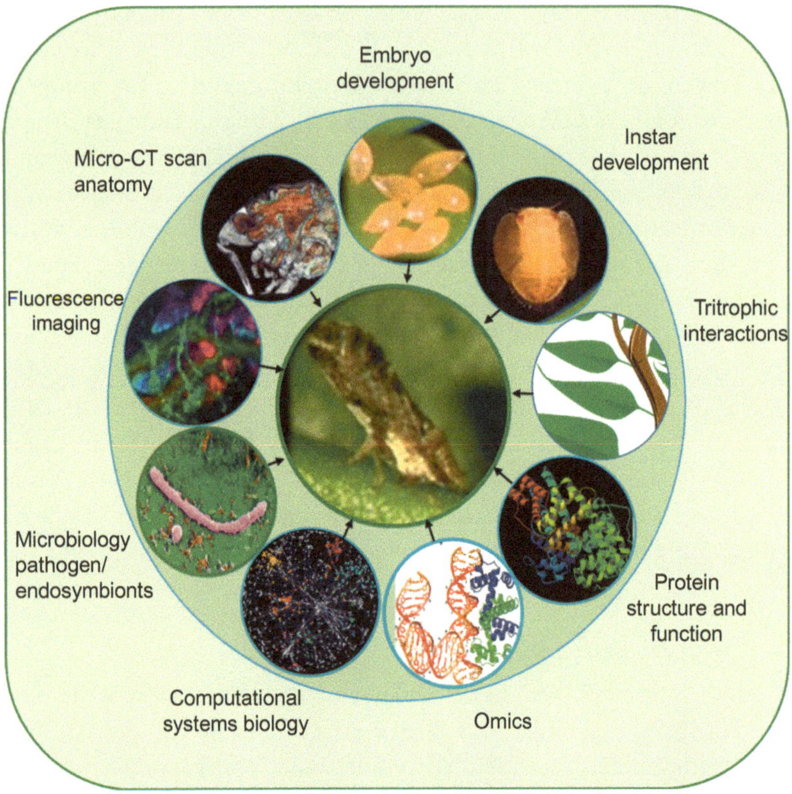

Fig. 1 Systems biology approach for deriving solutions from a comprehensive arthropod vector pathosystem model. Methods of interest: insect biology, pathology, and -omics studies (transcriptomes, genomes, proteome, metabolome, for all organisms within the ecosystem surrounding the pathosystem of interest). The interactions among microbes within each organism. Production of Big Data and development of applications, along with emerging technologies, will continue to increase the interactions between the community of researcher and public. Portals and instant access devices now include greater visualization of information. Data can often be better understood when including fluorescence imaging, digital video libraries such as the emerging technology of micro-computed tomography (micro-CT and nano-CT), combined with 3D imaging software and printing to bring anatomical structures to life (Alba-Tercedor et al., 2017; Cicero et al., 2018). More digital videos of psyllid anatomy available at www.citrusgreening.org.

2. PROGRESS IN PSYLLID GENE ANNOTATION AND RESOURCES

The basis to formulate appropriate genetic base solutions to the citrus greening pathosystem is the availability of high-quality genomic data from citrus, the psyllid, the bacterial pathogen, and microbes within the system, such as psyllid endosymbionts and tree endophytes. Endophytic bacteria and endosymbionts along with the gut fauna are able to lessen or prevent the deleterious effects of certain pathogenic organisms (Arp et al., 2016, 2017; Dossi et al., 2014; Eleftherianos et al., 2013; Ramsey et al., 2015).

2.1 Genome ACP

Public release of an improved ACP genome, DIACI_2.0, and Official Gene Set, DIACI_OGS_v2 (Saha et al., 2017a,b), was possible in part with support from National Institute of Food and Agriculture, USDA, Specialty Crops Research Initiative/Citrus Disease Research & Extension, Award #2015-70016-23028, Developing an Infrastructure and Product Test Pipeline to Deliver Novel Therapies for Citrus Greening Disease, Lead Dr Susan Brown, Kansas State University, Manhattan, KS, & USDA, ARS, Fort Pierce, FL. Data are available at www.citrusgreening.org and https://citrusgreening.org/organism/Diaphorina_citri/genome.

2.2 Endosymbiont Genomes in ACP

The psyllid, *D. citri*, harbours three endosymbiotic bacteria, *Candidatus* Carsonella ruddii, *C.* Profftella armatura, and *Wolbachia-Diaphorina*, wDi (Chu et al., 2016; Meyer and Hoy, 2008; Saha et al., 2012; Subandiyah et al., 2000). The facultative endosymbiont, *C.*P. armatura, has been proposed to produce antimicrobial polyketides, which may increase defences against harmful environmental microbes (Nakabachi et al., 2006, 2013; Ramsey et al., 2015). *Wolbachia* is an endosymbiont broadly associated with psyllids which can elevate the immune response and protect against viral pathogens (Pan et al., 2012; Ye et al., 2013). The genomes of the *D. citri* endosymbionts and *Wolb-Dc* (Saha et al., 2012) encompass 18 genome isolates from the five Liberibacter species [Las (8 genomes), Laf (1 genome), Lam (2 genomes)], which are available at www.citrusgreening.org and the NCBI database. Wang et al. (2017a) reported that potential CRISPR (clustered regularly interspaced short palindromic repeats) in most of the 18 sequenced Liberibacter genomes have been identified suggesting that a CRISPR interference strategies may work as a management tool. CRISPR interference (CRISPRi) is a genetic perturbation technique that allows for sequence-specific repression or activation of gene expression in prokaryotic and eukaryotic cells (Jiang et al., 2013; Li et al., 2016; Peters et al., 2016). It was first developed in the Lim and Weissman labs at UCSF (Qi et al., 2013) and has been used successfully to suppress mycobacteria (Choudhary et al., 2015).

2.3 Citrus Genomes

Finally, several citrus species genome datasets are available at www.citrusgreening.org for open access research. A review of *Citrus* as a genus, and the taxonomic classification of species has recently been reviewed and provides an excellent insight into the origins and development of *Citrus* we know today (Wu et al., 2018). Whole-genome shotgun-sequencing data is available at NCBI under *Citrus* BioProject PRJNA414519. At this time there is a wealth of genomic data and other biological information available on the organisms within the HLB pathosystem that should expedite research towards solutions to the ACP and citrus greening disease problem.

3. PROGRESS IN RNAI IN PSYLLIDS

3.1 Delivery of dsRNA Is Important for Successful RNAi

Delivery and response to RNAi across arthropods have generated a large amount of data and have demonstrated a wide range of sensitivity to RNAi treatments (Darrington et al., 2017; Thakur et al., 2016; Zotti et al., 2018). Since the RNAi effects of flower colour suppression were observed in the early 1990s (Napoli et al., 1990), it would not be until 1998 that the elucidation of the dsRNA trigger would be thoroughly described, and the term RNA interference coined (Fire et al., 1998) (for a historical overview, see Sen and Blau, 2006). The applications of RNAi technology in the management of pests and pathogens were quickly realized (Agrawal et al., 2003; Baum et al., 2007; Elbashir et al., 2001; Hannon, 2002; Huvenne and Smagghe, 2010; Karl et al., 2007; Price and Gatehouse, 2008; Turner et al., 2006).

The strategies and delivery methods of practical RNAi utility have previously been reviewed (Darrington et al., 2017; Ghosh et al., 2018; Ivashuta et al., 2015; Joga et al., 2016; Kanakala and Ghanim, 2016; Kolliopoulou et al., 2017; NASEM, 2016; Rodrigues and Figueira, 2016; Thakur et al., 2016; Zotti and Smagghe, 2015; Zotti et al., 2018). A variety of strategies can successfully be used for RNAi including: plant transformation (Hajeri et al., 2014; Lombardo et al., 2016; Saurabh et al., 2014); topical spray applications to plants or insects (Faustinelli et al., 2018; Koch et al., 2016; San Miguel and Scott, 2016); soil treatments as liquids or using natural carriers like clays and other absorbents (Ghosh et al., 2018; Mitter et al., 2017). Microbe expression of dsRNA in planta using: viruses, bacteria, yeast, or fungi (Kolliopoulou et al., 2017; Whitten and Dyson, 2017; Whitten et al., 2016; Zotti et al., 2018).

Analyses of RNAi-related genes in the ACP genome highlighted the genes representing the core RNAi pathway of psyllids also in other insects (Saha et al., 2017a; Taning et al., 2016). In arthropods, the core RNAi machinery consists of Argonaute endonucleases, Dicer enzymes, and dsRNA binding proteins. In insects there are three different RNAi pathways based upon the types of Dicers or Argonautes involved and the small RNAs (see reviews: Dowling et al., 2016; Taning et al., 2016). Like other insects, psyllids lack an RdRp and R2D2 sequences. The *r2d2*, dsRNA-binding protein, in *Drosophila* functions as a cofactor of Dicer-2, which is essential for both RNAi and innate immune responses against RNA viruses (Taning et al., 2016). Nevertheless, a variety of bioassays has demonstrated significant RNAi response in psyllids. Bioassays using in planta feeding systems (Andrade and Hunter, 2016, 2017; Galdeano et al., 2017; Ghosh et al., 2017, 2018; Kruse et al., 2017; Taning et al., 2016); detached leaf bioassays (Ammar et al., 2013; Raiol-Junior et al., 2017); leaf disc, artificial diet, sugar solutions (Russell and Pelz-Stelinski, 2015); topical applications, soaking (Killiny and Kishk, 2017; Killiny et al., 2014; Yu et al., 2017) all have shown a strong RNAi response across all the Asian psyllid life stages (El-Shesheny et al., 2013; Galdeano et al., 2017; Kishk et al., 2017). So if ACP is so RNAi responsive, why is so much variation in trials reported? Two major factors are proposed to be responsible for the variability observed in RNAi efficiency across arthropods: first, the presence of dsRNases during digestion and subsequent processing into siRNAs in the cells, and movement between cells (Singh et al., 2017), and second, the high variability in intercellular transport which has a strong bearing on dsRNA concentration and systemic RNAi efficiency (Burand and Hunter, 2013; Scott et al., 2013; Whitten and Dyson, 2017). With these concerns in mind, several experiments were conducted to improve the delivery, intercellular movement postdelivery (concentration of the trigger), and RNAi activity of ingested dsRNA in ACP.

Traceability of dsRNA to study delivery of dsRNA or PPMO in tissues of ACP and citrus leaves was accomplished by using the ULYSIS®

Nucleic Acid Labeling technology complex (owned by KREATECH Diagnostics) enabling rapid and simple coupling of the Alexa Fluor® dyes to purine bases in nucleic acid polymers. The method is built upon the versatile, patented Universal Linkage System (ULS) platinum-based chemistry (KREATECH Biotechnology BV). The platinum dye forms a stable adduct with the N-7 position of guanine, and, to a lesser extent, adenine bases in DNA, RNA, PNA, and oligonucleotides. The result is a reliable nonenzymatic method for nucleic acid labelling that takes about 15 min to complete. Standard analyses using PCR were also possible postlabelling (see Thermo Fisher Scientific; http://www.thermofisher.com/us/en/home/references/molecular-probes-the-handbook/nucleic-acid-detection-and-genomics-technology/labeling-oligonucleotides-and-nucleic-acids.html#head5). The dsRNAs evaluated were of lengths between 140 and 248 nt. The delivered G-labelled dsRNA was applied as a tagged control and as active trigger in RNAi trials. For soil application, a solution containing 100 mL water, pH 7.4–8.0, supplemented with 10 or 100 μg dsRNA total concentration was applied. The solution was poured onto the soil of potted citrus seedlings (1 gal. pots, 2.85 L); the roots absorbed the G-labelled dsRNA which was detected in the xylem and phloem of new growth leaves on treated citrus seedlings at 3 days postsoil applied treatments. These results support previous observations on larger citrus trees and grapevines, in which the dsRNA was detectable for over 45 days postsoil applications (Hunter et al., 2012). Seedlings were 2-year-old potted sweet orange *Citrus sinensis* (L.) Osbeck. Leaf tests are described in Fig. 2.

4. IMPROVING PERSISTENCE AND ACTIVITY (USE OF NONCANONICAL NUCLEOSIDES)

4.1 Noncanonical dsRNA to Increase Persistence and RNAi Activity

Experiments demonstrated downregulation of *arginine kinase* in ACP, through ingestion of an *ak*-dsRNA, of 200 nt in length, fed through in planta bioassay (Andrade and Hunter, 2017) which caused low levels of mortality (~18%) (Fig. 3) (Andrade and Hunter, 2017; Clarke et al., 2018; Taning et al., 2016). Concentrations of 1 and 2 μg canonical *ak*-dsRNA (MEGAscript T7 kit, Thermo Fisher Scientific, according to manufacturer's instructions) when fed to ACP resulted in mortalities of 4% and 6% greater, respectively, than the negative control Chinese sacbrood virus, CSBV-dsRNA

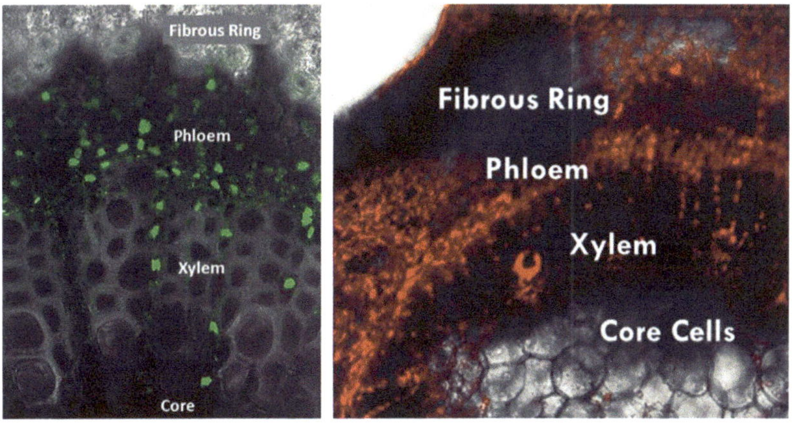

Fig. 2 Fluorescent-labelled dsRNA shows systemic movement in citrus trees and leaves, sweet orange, *Madam vinous* and *Citrus sinensis*. Soil-applied and root-absorbed labelled dsRNA was detected in the xylem and phloem of new growth leaves on treated citrus seedlings at 3 days postsoil applied treatments. The results supported previous results on larger citrus trees and grapevines in the field which detected dsRNA for over 45 days postsoil applications to the root zones (Hunter et al., 2012). Seedlings were 2-year-old potted sweet orange. *Citrus sinensis* (L.) Osbeck. Leaf tests: the control leaf (*left*) was a check for autofluorescence, leaf counterstained with *NucGreen*® Dead 488 reagent, which is suitable for staining nuclei in fixed cell preparations and tissue sections (Thermo Fisher Scientific). Treated citrus leaf (*right*). Mid rib cross section, at 72 h postabsorption with Fluoro-tagged dsRNA (ULYSIS® Nucleic Acid Labeling Kit; Thermo Fisher Scientific): Molecular Probes, Cat. No. U21650 (Hunter et al., 2018c).

on day 8 (Fig. 3, orange bars). However, incorporation of noncanonical dsRNA (49.6% C & U, replacements, DuraScribe® T7 Transcription Kit) fed at the same concentrations of 1 and 2 μg (red bars) resulted in a significant increase in ACP mortalities that were 12% and 18% greater, respectively, compared to the canonical *ak*-dsRNA fed and CSBV-dsRNA control at 8 days postfeeding access (Hunter and Lopez, 2018; Hunter et al., 2018c) (Fig. 3).

4.2 Canonical and Noncanonical *ak*-dsRNA Knockdown in ACP

Both canonical and noncanonical *ak*-dsRNA reduced arginine kinase in ACP. A 1.26-fold suppression was observed in psyllids when fed on citrus cuttings treated with 1 μg *ak*-dsRNA, and a 1.31-fold suppression when fed on cuttings with 2 μg dsRNA. Additionally, the noncanonical dsRNA resulted in a 1.1-fold knockdown when fed with 1 μg and a 1.6-fold knockdown when fed on treatments with 2 μg noncanonical dsRNA. The incorporation of alterations of the components in oligos will continue to provide

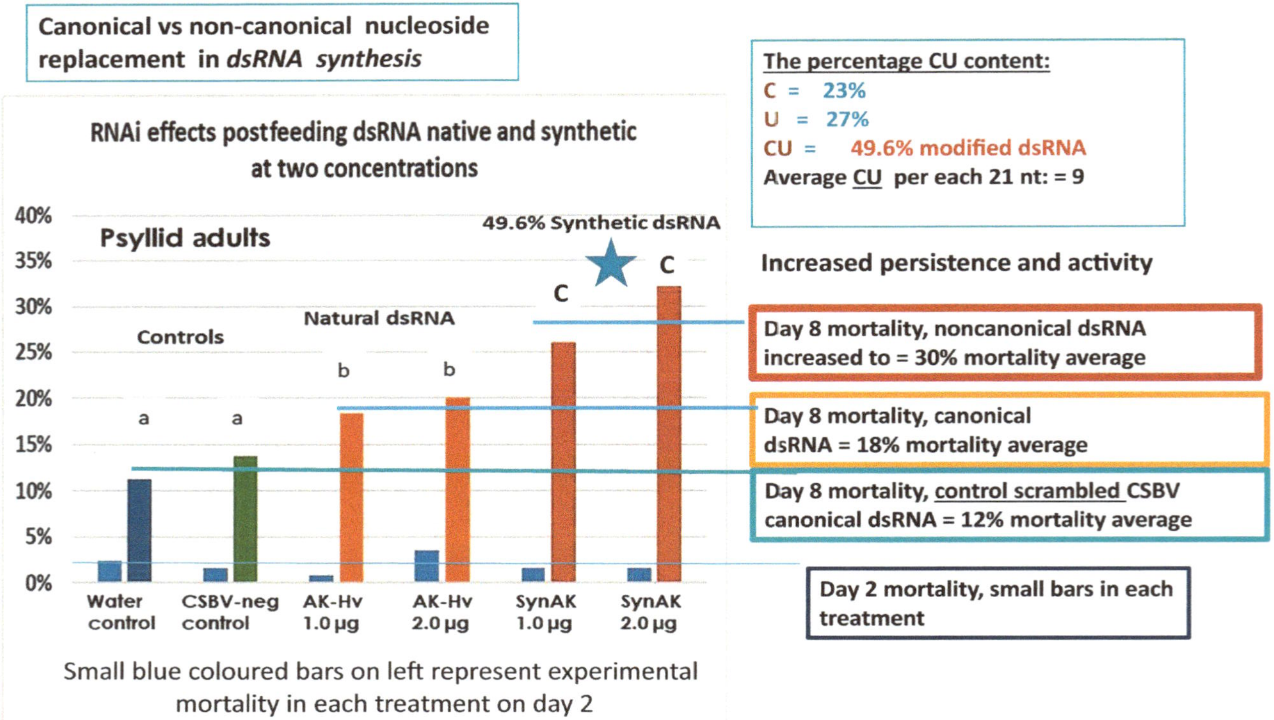

Fig. 3 Increases in mortality and transcript suppression resulted when half the nucleosides in the dsRNA were substituted with noncanonical forms for 'C' and 'U'. Comparison between the same dsRNA sequence made with two different kits. The traditional MegaScript® kit using canonical nucleosides, compared to dsRNA made with replacement of the 'C' and 'U' with noncanonical nucleosides (Hunter and Lopez, unpublished). Produced using the DuraScribe® T7 Transcription Kit (Cat. No. DS010910 & DS010925, www.Lucigen.com).

increased benefits in stability, cell entry, sequence binding specificity, and activity. Like most emerging technology, the cost of production will continue to drop and use of these products will then continue to increase.

Rapid degradation is one of the problems with dsRNA applied as topical sprays or when orally ingested (Allen and Walker, 2012; Dubelman et al., 2014; Terenius et al., 2011). The use of altered or noncanonical nucleosides greatly improves RNAi applications of exogenous dsRNA. The use of synthetic nucleosides, even in single-stranded RNA, ssRNA, complimentary to the mRNA provides increased efficacy (Hegarty and Stewart, 2018; Ludin et al., 2015; Shen and Corey, 2018). The growing use of antisense oligos, which have a long history of modifications of nucleotides, and sugars continues to demonstrate the beneficial properties of modifications to improve efficacy (Bennett and Swayze, 2010; Crooke, 2017; Hegarty and Stewart, 2018; Juliano, 2016; Kurreck, 2003; Ludin et al., 2015; Shen and Corey, 2018; Verma, 2018).

5. ASSISTED DELIVERY MECHANISMS

In order to improve delivery and activity, and to prevent undesirable degradation, biotic and abiotic delivery systems are being developed. These systems include the use of lyposomes, chitosan nanoparticles, bacteria, yeast, and viral expression systems that have been explored for applications in arthropods (Hajeri et al., 2014; Murphy et al., 2016; Mysore et al., 2014; Taning et al., 2016; Whyard et al., 2015; Zhang et al., 2010; Zotti et al., 2018). One promising delivery strategy is to use cationic lipids, polymers, dendrimers, peptides, or organic nanoparticles, which permit more modifications for target delivery (Darrington et al., 2017; Gillet et al., 2017). Therapeutic delivery systems for medical treatments are estimated to exceed a US$30 billion annual market (Wong et al., 2017). Use of virus expression systems have previously been reviewed (see Kolliopoulou et al., 2017). Advances in inert carriers, or directed carriers to specific cell types, continue to improve and will soon be able to target specific cells in the organism of choice. These types of advances into agriculture settings come from rapid development in medicine and cancer treatments (Barros et al., 2016; Mout et al., 2017; Sinisterra-Hunter and Hunter, 2018; Wong et al., 2017). These same strategies will rapidly be incorporated into agricultural settings due to fewer regulatory hurdles. As delivery issues are resolved, the efficacy of RNA suppression technologies will increase across more arthropod systems. In insects which already show high RNAi activity, such as in ACP, the adoption of these technologies may help companies to commercialize RNA-suppressing products (Hunter et al., 2017; Sinisterra-Hunter and Hunter, 2018).

5.1 Ribonucleoprotein Particle

A recent report using PTD-DRBD, peptide transduction domain-dsRNA binding domain, by Gillet et al. (2017) showed improved RNAi efficacy in *Anthonomus grandis*, the cotton boll weevil, postingestion by using the complex PTD-DRBD:dsRNA to enhance gene silencing. The ribonucleoprotein particle (RNP) was capable of entering cells via endocytosis and escaping the endosome thus delivering the dsRNA into the cell cytoplasm providing improved triggering of RNAi to silence the targeted gene (Eguchi et al., 2009; Wadia et al., 2004).

5.2 Branched Amphiphilic Peptide Capsules

Branched amphiphilic peptides self-assemble into capsules or vesicles (BAPC). The peptide nanospheres are comprised of equimolar proportions of two branched peptide sequences bis(FLIVI)-K-KKKK and bis(FLIVIGSII)-K-KKKK that self-assemble to form bilayer delimited capsules. In two publications the specific lipid analogous characteristics of BAPC were described in detail (Sukthankar et al., 2014; Tomich et al., 2014). When incorporated into RNAi studies, Avila et al. (2018) reported a significant increase in delivery and response to lethal dsRNAs when mixed into insect diets incorporating BAPC to *Tribolium castaneum* (Coleoptera), and the pea aphid, *Acyrthosiphon pisum* (Hemiptera). Improved delivery combined with slower controlled release provided increased RNAi efficacy. Emergence of nanomaterials as carriers for gene-targeting treatments for medical or agricultural needs is a rapidly growing field of development. The use of BAPC, an inert peptide, to deliver dsRNA to two different orders (Coleoptera and Hemiptera) of insects: The *Tribolium castaneum*, flour beetle, Coleoptera, on a solid diet, and the pea aphid, *Acyrthosiphon pisum*, Hemiptera, on a liquid diet, has been demonstrated (Avila et al., 2018). In their study, they describe the use of a peptide nanomaterial, branched amphiphilic peptide capsules (BAPC), that facilitates cellular uptake of dsRNA by insects through feeding. The insect gene transcript used were BiP and Armet. These transcripts are part of the unfolded protein response (UPR) (Jarvis et al., 1990; Wang et al., 2015). Suppression of their translation resulted in increased mortality. Ingestion of BiP-dsRNA by

Acyrthosiphon pisum when associated with BAPC led to an LC50 of 4–5 days compared to controls that had ingested the same amounts of BiP-dsRNA without BAPC supplementation which resulted in a LC50 of 11–12 days. Ingestion of a combination of BiP-dsRNA plus Armet-dsRNA complexed with BAPC fed to *Tribolium castaneum* resulted in greater mortality with most dying as larvae or during eclosion (~75%). Feeding the combined dsRNAs alone, no BAPC, produced significantly less mortality (~30%). Feeding of the BAPC with no dsRNA to either insect did not result in significant differences from the blank no dsRNA, water controls. The results show that complexation of dsRNA with BAPC enhanced the oral delivery of dsRNA over dsRNA alone. Preliminary trials with BAPC and increased RNAi efficacy with ACP also result in increased mRNA target suppression (Hunter, unpublished). The sensitive and effective response by ACP to ingested dsRNA to significantly suppress mRNA suggests that an RNAi strategy, whether produced by the plant, a microbe, or exogenously applied as foliar or soil treatments, should aid efforts to reduce ACP without negative effects on nontarget species (Andrade and Hunter, 2016).

6. ANTISENSE OLIGONUCLEOTIDES

6.1 Morpholinos

Since the naked Morpholino backbone has no charge, a Morpholino will not produce naked Morpholino backbone has no charge. A Morpholino will not produce the charge–charge interactions needed upon arriving at the active site of a nucleic acid binding protein. The lack of sufficient force between the hydrogen bonds is too weak to hold a Morpholino in the catalytic site of a nuclease or bind it into a Toll-like receptor (Moulton, 2016). Morpholinos must be covalently linked with a charged molecule or peptide, such as a cell-penetrating peptide, CPP, to enter bacteria (both Gram-positive or Gram-negative). After the MOs are conjugated to a carrier, they are referred to as PPMO, CPPMO, PMO, or similar (Moulton, 2016; Moulton et al., 2004; Rasmussen et al., 2007; Zatsepin et al., 2005). For more protocols, see: Moulton and Moulton (2017), with other references at GeneTools, LLC, website (http://www.gene-tools.com/morpholino_antisense_oligos).

Morpholinos have little to no interaction with nucleic acid–binding proteins, are very stable, and thus far there are no reported enzymes, which degrade Morpholinos (Hudziak et al., 1996; Youngblood et al., 2007). The original designs of Morpholinos were as steric blockers, which would lock onto the RNA and prevent translation into protein (Summerton and Weller, 1997). Short PPMOs of 14–18 nt are used to target bacteria, and the sequence, referred to as the external guide sequence, EGS, when complexed with the target RNA, modifies the shape to resemble a tRNA precursor which renders the target RNA susceptible to cleavage by ribonuclease P (RNase-P) (Altman and Kirsebom, 1999). RNase-P is an enzyme in all cells, which cleaves tRNA precursors to generate the 5′ termini of mature tRNAs (Altman, 1989; Forster and Altman, 1990; Jarrous, 2017; Yuan and Altman, 1994). Altman (2011) envisioned the ability to target any RNA or mRNA in cells which have RNase-P, which would enable targeting and eliminating drug-resistant bacteria (Moulton, 2016; Moulton and Yan, 2008; Sala et al., 2012).

One example of using a Morpholino to treat fruit trees, which has some similarity to the Liberibacter pathosystem in citrus, is the suppression of the pathogen *Erwinia amylovora*, a Gram-negative bacterial plant pathogen (Family: Enterobacteriaceae). The *Erwinia* bacteria are the causal agent of fire blight. This pathogen is the most serious disease of apple and pear in the USA and worldwide. The infection affects all parts of the tree, flowers, leaves, and fruits resulting in rapid yield loss and eventual tree death (Norelli et al., 2003; van der Zwet et al., 2012). The current treatment is the application of the antibiotic streptomycin during bloom. Annual production losses are over US$100 million annually in the USA (Norelli et al., 2003). Of great concern after many years of treatment with antibiotics, the fire blight bacteria are developing resistance to streptomycin (McGhee et al., 2011; Patel et al., 2017). Although less effective than streptomycin, the anti-acpP-CPP1 at 100 mM when sprayed on flowers was able to effectively limit the pathogen growth on the stigmas of apple flowers. The research also indicated that efficacy was linked to CPP conjugation with the PNA, before significant antimicrobial effects were induced in *E. amylovora* (Patel et al., 2017).

Similarly, Hunter and Pelz-Stelinski proposed to eliminate Las from infected citrus trees, and ACP using PPMOs, and to reduce ACP by suppressing their endosymbionts (Altman, 2013). The proposed research included the development of Morpholino-based bactericides to reduce pathogen transmission by ACP, while also reducing or eliminating infections in existing infected citrus trees (Pelz-Stelinski et al., 2017). Trials using PPMOs to suppress bacteria in cultures were successful and supported previous reports by Wesolowski et al. (2013). The cell culture bioassay used a

PPMO provided by Dr Sid Altman, Yale University, which was designed to a region of *gyrase A* in *E. coli*, known to inactivate several bacterial species. Their research showed that the binding affinity was dependent upon the number of SNPs in the 18nt binding sequence of the PPMO (Wesolowski et al., 2011, 2013). Comparison of the *gyrase A* region in the PPMO designed to *E. coli*, to the same regions of interest: Las, Profftella armatura, and *Wolbachia-Diaphorina*, wDi, showed that it had 100% identical match to the wDi *gyrase A* gene. The PPMO was reported to decrease the wDi *Gyrase* target (Pelz-Stelinski et al., 2017; Sandoval-Mojica et al., 2017). In trials on PPMO movement through plants, citrus cuttings, and whole seedlings, the PPMO was labelled with Lissamine fluorophore. The PPMO was visualized systemically in all plants tested (citrus seedlings, *Murraya paniculata*, okra—*Abelmoschus esculentus*, periwinkle—*Vinca minor*, and sunflower—*Helianthus annuus*, and potato) (Hunter et al., 2018a). When ACP was provided feeding access on treated plants, the PPMO tag was also visualized using confocal microscopy within psyllid tissues (Hunter et al., 2018a,b). Root absorption was validated by two different methods: direct soil application (Hunter) and root dipping (Pelz-Stelinski). Both methodologies were equally efficient in the delivery of PPMO into plants (Hunter and Lopez, 2018; Hunter et al., 2018a; Pelz-Stelinski et al., 2017). A protocol for designing PPMO bacterial knockdowns was previously developed as a potential alternative to antibiotics (Daly et al., 2017). The history of other landmark discoveries in oligonucleotide research is outlined by Ludin et al., (2015). Development for medical use has been previously reviewed (Hegarty and Stewart, 2018; Shen and Corey, 2018). For information on improved designs that increase binding affinity, resistance to nucleases, and function, see reviews: Bennett and Swayze, 2010; Hagedorn et al., 2018; Ludin et al., 2015).

Research on the need for transfection agents or additional carriers proteins, nanoparticles, chemicals), have made significant advances to improve product delivery into cells and specific tissues (Bennett and Swayze, 2010; Hagedorn et al., 2018; Hegarty and Stewart, 2018; Ludin et al., 2015; Shen and Corey, 2018). Peptides are also being used to improve delivery of CRISPR/Cas9 (Chaverra-Rodriguez et al., 2018; Marshall and Akbari, 2018), and other geneediting technologies (e.g. TALENS, CRISPR-Cas9) (Hagedorn et al., 2018; Ludin et al., 2015; Wang et al., 2016).

Morpholino's, like PPMO, when tagged with a fluorophore could be detected in plant tissues when absorbed by roots or cuttings (citrus, grapevines, potato, periwinkle, tomato), and when injected directly into the plant stem or tree trunk (Hunter et al., 2018a,b). The fluorescence-tagged PPMO was visualized using confocal microscopy in plant leaves and insect tissues, and by use of a Synergy H1 fluorescence plate reader (Hunter et al., 2018a,b). A competitive binding assay also demonstrated efficiency as related to the presence of SNPs in the target region of the RNA (Hunter et al., 2018a; Wesolowski et al., 2013). PPMO are highly resistant to degradation (Hudziak et al., 1996; Youngblood et al., 2007), which provides longer persistence than naked dsRNA applied in the similar fashion (Hunter et al., 2012; Dubelman et al, 2014; Moulton and Yan 2008) lasting from weeks to months, as opposed to days in treated citrus trees (Andrade and Hunter 2017; Clarke et al., 2018; Ghosh et al., 2018; Hunter et al., 2012; 2017).

6.2 PPMO Targeting Bacteria in Citrus Trees and ACP

Bacteria are difficult to eliminate from a living system. The fast rates of replication and protection under biofilms make them a formidable organism to treat. Treatments with PPMO were able to reduce Las bacteria in infected citrus seedling trees, *Madam vinous*, or *Citrus sinensis*. Direct tree injectors with 10 mL volumes were used, and the 3-year-old citrus seedlings were kept in the glasshouse at ambient light and temperature (Pelz-Stelinski et al., 2017). The potted Las-infected citrus seedling trees were produced at the USDA, ARS, US Horticultural Research Lab, Fort Pierce, FL (Dr Greg McCollum). These trials were the first reported evidence that PPMO have the potential as treatments to reduce Liberibacter bacteria in infected citrus trees. While the initial reports used single-dose treatments, trials underway will focus on multiple-dose treatments to determine the required concentration to keep Las concentrations suppressed.

7. GENE EDITING STRATEGIES IN ACP

7.1 BAPC-CRISPR/CAS9 System, Knockout in ACP

Biotechnology has produced a range of methods which permit gene regulation, modification, and editing that now includes all living organisms from bacteria, plants, animals, and humans. Clustered regularly interspaced short palindromic repeats (CRISPR) and the CRISPR-associated gene Cas9 represent an invaluable system for the precise editing of genes across all species

(Doudna and Charpentier, 2014; Wang et al., 2016). CRISPR/Cas9 is a bacteria system of defence with an adaptive mechanism that enables bacteria and archaeal species to resist invading viruses and plasmids (Doudna and Charpentier, 2014; Garneau et al., 2010). Due to the relative ease of use and increased efficiency CRISPR methods have become the primary gene-editing tool across the life sciences. Extensive information on the description and mechanism of action of CRISPR and other gene editing mechanisms such as zinc finger nucleases, ZFN, and TAL effector nucleases is provided in the following reviews: TALENS (Bortesi and Fischer, 2015; Gaj et al., 2013; Markert et al., 2016; Sun et al., 2017), and clustered regularly interspaced short palindromic repeats (CRISPR) and the CRISPR-associated proteins (Cas), referred to as: (CRISPR-Cas9) system (Boettcher and McManus, 2015; Dominguez et al., 2016; Gupta and Shukla, 2016; La Russa and Qi, 2015; Liang et al., 2015; Wang et al., 2016; Wilson and Doudna, 2013). CRISPR/Cas9 continues to be used successfully in a growing number of insects and arthropods (Champer et al., 2016; Chen et al., 2016, 2017; Cui et al., 2017; Gantz and Akbari, 2018; Sun et al., 2017; Taning et al., 2017). CRISPR-Cas9 was used to conducting a knockout, in ACP, of the Thioredoxin gene. *Diaphorina citri* is predicted to have at least two thioredoxins, TRX-1 and TRX-2, with variants in the mitochondria and cytoplasm. To prepare the guide RNA and determine the target DNA with a proper protospacer adjacent motif, purchase of the CRISPR-associated protein 9 (Cas9) and numerous publications on CRISPR protocols (Bassett et al., 2013; Larson et al., 2013; Zhang and Reed, 2017), YouTube videos, and discussions with researchers who have had success were all sought and consulted (Garczynski et al., 2017). The guides were made to the predicted: *Diaphorina citri* thioredoxin-2-like (LOC103521994) gene, for transcript variant X1 (sequence ID: XM_008487100.1) the fragment has a length of 1243 bp as described in genomic assembly DIACI_v2.0: https://citrusgreening.org/organism/Diaphorina_citri/genome. The production of the multiple sgRNAs was performed by Dharmacon Inc. (Lafayette, CO) (http://dharmacon.horizondiscovery.com/gene-editing/crispr-cas9/#all). The psyllid TRX-2 was selected for knockout, targeting a deletion of 556 nt. Trials of 30 nymphs (4th and 5th instar), and 20 adult females per treatment trials received coinjection of two sgRNAs (100 ng/ μL of each, with 200 ng/μL of Cas9 protein) (Zhang and Reed, 2017), while laying on chilled agarose gel (1%). Postinjections the psyllids were transferred to a citrus seedling tree for each group, to mate and oviposition. Seven days posttreatments a cohort of six adults were individually analysed using primers to the gDNA bracketing the predicted deletion at 200–250 beyond the ends in each direction. The

BAPC-CRISPR/Cas9-treated psyllids resulted in a delay in the number of days to eclose to adults (average of 8 days longer compared to controls injected with buffer). Average mortality posttreatments was 58% in both treatments. Adult longevity and fecundity were reduced in TRX psyllids, with an average longevity of 8.5 days posteclosion, vs controls which averaged 16 days. Thioredoxin participates in various redox reactions through the reversible oxidation of its active centre dithiol to a disulphide and catalyses dithiol–disulphide exchange reactions. As a reducing substrate of peroxiredoxin 1, thioredoxin 2 is preferred over thioredoxin 1, and required for female meiosis and early embryonic development. TRX regulates the functions of at least 30 target proteins, some of which are enzymes and transcription factors. TRX also plays an important role in the defence against oxidative stress by directly reducing hydrogen peroxide and certain radicals, and by serving as a reductant for peroxiredoxins, is involved not only in cytoprotective functions against oxidative stress, but also in the regulation of cellular proliferation and the ageing process (Yoshida et al., 2005).

Highly specific, biotechnologies, like RNAi and CRISPR, provide new sustainable and environmentally friendly strategies to reduce psyllids and many other insect vectors. Adult and nymphal psyllids were reduced postingestion of double-stranded RNA, dsRNA. A new system using BAPC-CRISPR/Cas9 injected into 4th, 5th instars, and adult females near ovaries, successfully resulted in heritable germline gene knockout, with deletions of 220 and 505 bp in the TRX gene. The psyllids with TRX KO had longer developmental time taking on average 6 days longer to eclosion into adults.

The TRX-treated psyllid adults lifespan averaged 8.5 days posteclosion compared to controls injected with buffer, which averaged 16 days. Average mortality was 58%–69% across treatments at 3 days postinjections.

The CRISPR knockout, KO, of TRX resulted in psyllids with: longer development times, shorter adult lifespans, and reduced fecundity. There is room for a lot of improvement with this technique.

Interestingly of the few F1 eggs oviposited, post-TRX injection near ovaries of an F0 adult female which also was KO positive, when collected as nymphs, one 4th instar out of six analysed was TRX KO positive. Surprisingly the development of a second-generation psyllid from eggs, NOT injected, but oviposited from a TRX-injected F0 adult female, had the missing TRX-KO sequence. Thus, it appears that (1) purposefully coinjecting Cas9 protein and two sgRNAs, with 0.1 ng BAPC, into the region of the adult female psyllid ovaries may produce stable heritable germline KO offspring (Hunter et al., 2018c). (2) A BAPC-CRISPR system would provide a much easier gene-editing strategy for insects where it has not been feasible to conduct egg injections. Previous egg injections trials in psyllids over a 1-year period, and thousands of eggs, did not produce any positive KO results, and few psyllids.

8. FUTURE OUTLOOK

8.1 Considerations and Outlook

The use of a multitude of gene-targeting strategies, RNAi, PPMO, and CRISPR/Cas9, has been successfully applied to reduce ACP survival. The evidence from just a few studies have shown the great potential across all the methods described as well as for many others which have yet to be tried.

RNAi has by far the greatest amount of research support as a strategy that works in psyllids and other hemipterans (Andrade and Hunter, 2016; Ghosh et al., 2018; Joga et al., 2016; Li et al., 2013, 2015). The current research has shown that RNAi can reduce ACP populations when the dsRNA molecules deliver both, as an exogenous product and as a result of genetic modification of the citrus trees. Beyond the RNAi technologies mentioned here, there are many other technologies which can be used to target the bacteria in the ACP/Liberibacter/Citrus greening disease pathosystem. Thus, efforts to rapidly advance these technologies, cell-penetrating peptide Morpholino, PPMO, CRISPR/Cas9, and CRISPRi, to target microbes in the psyllid and in citrus trees need more research which is focused on developing partnerships with commercial companies with an interest to move towards commercialization.

ACKNOWLEDGEMENTS

The authors wish to acknowledge the contributions by Dr Kirsten Pelz-Stelinski, University of Florida, for PPMO targeting psyllid microbes. We thank Maria T. Gonzalez, Salvador P. Lopez, Jennifer Wildonger, Biological Science Technicians, bioassays, qPCR analyses, imaging, USDA, ARS, Fort Pierce, FL. Jackie L. Metz, University of Florida, Ft. Pierce, FL. The LabEx/USA International program: Brazil/USDA, ARS, Dr Eduardo C. Andrade, EMBRAPA Cassava and Fruits, Rua Embrapa, s/n, Cruz das Almas, Bahia, Cep 44380-000, Brazil, cooperative research program on RNAi to reduce insect pests of citrus, 2011–2015. Michal Boyle, Smithsonian Marine Station, Confocal microscopy, and Chris S. Holland, confocal imaging, bioassay trials, ORISE program DOE/USDA. This research was supported in part by an appointment to the Agricultural Research Service (ARS) Research Participation Program administered by the Oak Ridge Institute for Science and Education (ORISE) through an interagency agreement between the US Department of Energy (DOE) and the US Department of Agriculture (USDA). ORISE is managed by ORAU under DOE contract number DE-SC0014664. All opinions expressed in this paper are the author's and do not necessarily reflect the policies and views of USDA, ARS, DOE, or ORAU/ORISE. Funding in part from: National Institute of Food and Agriculture, USDA, Specialty Crops Research Initiative/Citrus Disease Research & Extension. Award #2015-70016-23028, Developing an Infrastructure and Product Test Pipeline to Deliver Novel Therapies for Citrus Greening Disease, Lead Dr Susan Brown. Kansas State Univ., Manhattan, KS, & USDA, ARS, Fort Pierce, FL; and the NIFA, USDA, 2015 National Institute of Food and Agriculture, USDA, Specialty Crops Research Initiative. Award #2015-10479. Targeting microbes to control huanglongbing disease of citrus #2016-70016-24782, Lead Dr Kirsten Pelz-Stelinski, University of Florida, Lake Alfred, FL; Dr Wayne Hunter, USDA, ARS, Ft. Pierce, FL; Dr Sidney Altman, Yale University, New Haven, CT.

Mention of trade names or commercial products in this publication is solely for the purpose of providing specific information and does not imply endorsement or recommendation by the US Department of Agriculture. USDA is an equal opportunity provider and employer.

REFERENCES

Agrawal, N., Dasaradhi, P.V., Mohammed, A., Malhotra, P., Bhatnagar, R.K., Mukherjee, S.K., 2003. RNA interference: biology, mechanism, and applications. Microbiol. Mol. Biol. Rev. 67, 657–685. https://doi.org/10.1128/MMBR.67.4.657-685.2003.

Alba-Tercedor, J., Hunter, W.B., Cicero, J.M., Bariain, M.S., Brown, S.J. 2017. Use of micro-CT to elucidate details of the anatomy and feeding of the Asian citrus psyllid *Diaphorina citri* Kuwayama, 1908 (Insecta: Hemiptera, Liviidae). Micro-CT User Meeting 2017. Pg. 270–285. Royal Belgian Institute of Natural Science, Brussels.

Allen, M.L., Walker, W.B., 2012. Saliva of *Lygus lineolaris* digests double stranded ribonucleic acids. J. Insect Physiol. 58, 391–396. https://doi.org/10.1016/j.jinsphys.2011.12.014.

Altman, S., 1989. Ribonuclease P: an enzyme with a catalytic RNA subunit. Adv. Enzymol. Relat. Areas Mol. Biol. 62, 1–36.

Altman, S., 2011. Ribonuclease P. Philos. Trans. R. Soc. Lond. B Biol. Sci. 366, 2936–2941. https://doi.org/10.1098/rstb.2011.0142.

Altman, S., 2013. Antibiotics present and future. FEBS Lett. 588, 1–2. https://doi.org/10.1016/j.febslet.2013.10.048.

Altman, S., Kirsebom, L., 1999. Ribonuclease P. In: The RNA World, second ed. Cold Spring Harbor Laboratory Press. 0-87969-561-7/99 (Chapter 14).

Ammar, E.D., Shatters, R.G., Hall, D.G., 2011a. Localization of Candidatus Liberibacter asiaticus, associated with citrus Huanglongbing disease, in its psyllid vector using fluorescence in situ hybridization. J. Phytopathol. 159 (11 – 12), 726–734. https://doi.org/10.1111/j.1439-0434.2011.01836.x.

Ammar, E., Shatters, R.G., Lynch, C., Hall, D.G., 2011b. Detection and relative titer of Candidatus Liberibacter asiaticus in the salivary glands and alimentary canal of *Diaphorina citri* (Hemiptera: Psyllidae) vector of citrus huanglongbing disease. Ann. Entomol. Soc. Am. 104 (3), 526–533. https://doi.org/10.1603/AN10134.

Ammar, E.-D., Walter, A.J., Hall, D.G., 2013. New excised-leaf assay method to test inoculativity of Asian citrus psyllid (Hemiptera: Psyllidae) with Candidatus Liberibacter asiaticus associated with citrus huanglongbing disease. J. Econ. Entomol. 106, 25–35.

Ammar, E.-D., Ramos, J.E., Hall, D.G., Dawson, W.O., Shatters Jr., R.G., 2016. Acquisition, replication and inoculation of Candidatus Liberibacter asiaticus following various acquisition periods on huanglongbing-infected citrus by nymphs and adults of the Asian citrus psyllid. PLoS One 11 (7), e0159594. https://doi.org/10.1371/journal.pone.0159594. PMID: 27441694.

Ammar, E.-D., Hall, D.G., Hosseinzadeh, S., Heck, M., 2018. The quest for a non-vector psyllid: natural variation in acquisition and transmission of the huanglongbing pathogen 'Candidatus Liberibacter asiaticus' by Asian citrus psyllid isofemale lines. PLoS One 13 (4), e0195804. https://doi.org/10.1371/journal.pone.0195804.
Andrade, E.C., Hunter, W.B. 2016. RNA interference—natural gene-based technology for highly specific pest control (HiSPeC). RNA Interference (ed. by IY Abdurakhmonov), pp. 391–409. InTech, Rijeka, Croatia (Chapter 19). https://doi.org/10.5772/61612.
Andrade, E.A., Hunter, W.B., 2017. RNAi feeding bioassay: development of a non-transgenic approach to control Asian citrus psyllid and other hemipterans. Entomol. Exp. Appl., 162, 389–396. https://doi.org/10.1111/eea.12544.
Arp, A.P., Pelz-Stelinski, K.S., Hunter, W.B., 2016. Annotation of the Asian citrus psyllid genome reveals a reduced innate immune system. Front. Physiol. 7, 570. https://doi.org/10.3389/fphys.2016.00570.
Arp, A.P., Martini, X., Pelz-Stelinski, K.S., 2017. Innate immune system capabilities of the Asian citrus psyllid, *Diaphorina citri*. J. Invertebr. Pathol. 148, 94–101. https://doi.org/10.1016/j.jip.2017.06.002.
Aubert, B., 1987. *Trioza erytreae* Del Guercio and *Diaphorina citri* Kuwayama (Homoptera: Psylloidea) the two vectors of citrus greening disease: biological aspects and possible control strategies. Fruits 42, 149–162.
Avery, P.B., Hunter, W.B., Hall, D.G., Jackson, M.A., Powell, C.A., Rogers, M.E., 2009. *Diaphorina citri* (Hemiptera: Psyllidae) infection and dissemination of the entomopathogenic fungus *Isaria fumosorosea* (Hypocreales: Cordycipitaceae) under laboratory conditions. Fla. Entomol. 92, 608–618.
Avery, P.B., Wekesa, V.W., Hunter, W.B., Hall, D.G., McKenzie, C.L., Osborne, L.S., Powell, C.A., Rogers, M.E., 2011. Effects of the fungus *Isaria fumosorosea* (Hypocreales: Cordycipitaceae) on reduced feeding and mortality of the Asian citrus psyllid, *Diaphorina citri* (Hemiptera: Psyllidae). Biocontrol. Sci. Technol. 21, 1065–1078.
Avila, L.A., Chandrasekar, R., Wilkinson, K.E., Balthazor, J., Heerman, M., Bechard, J., Brown, S., Park, Y., Dhar, S., Reeck, G.R., Tomich, J.M., 2018. Delivery of lethal dsRNAs in insect diets by branched amphiphilic peptide capsules. J. Control. Release 273, 139–146. https://doi.org/10.1016/j.jconrel.2018.01.010.
Barros, W.M., Whitaker, W.K., Sukthankar, P., et al., 2016. A review of solute encapsulating nanoparticles used as delivery systems with emphasis on branched amphipathic peptide capsules. Arch. Biochem. Biophys. 596, 22e42. https://doi.org/10.1016/j.abb.2016.02.027.
Bassett, A.R., Tibbit, C., Ponting, C.P., Liu, J., 2013. Mutagenesis and homologous recombination in Drosophila cell lines using CRISPR/Cas9. Biol. Open 3, 42–49.
Baum, J.A., Roberts, J.K., 2014. Progress towards RNAi-mediated insect pest management. In: Advances in Insect Physiology, first ed. Elsevier Ltd.
Baum, J., Bogaert, T., Clinton, W., Heck, G.R., Feldmann, P., Ilagan, O., et al., 2007. Control of coleopteran insect pests through RNA interference. Nat. Biotechnol. 25, 1322–1326. https://doi.org/10.1038/nbt1359. 17982443.
Bawa, A.S., Anilakumar, K.R., 2013. Genetically modified foods: safety, risks and public concerns—a review. J. Food Sci. Technol. 50 (6), 1035–1046. https://doi.org/10.1007/s13197-012-0899-1.
Bellés, X., 2010. Beyond *Drosophila*: RNAi in vivo and functional genomics in insects. Annu. Rev. Entomol. 55, 111–128.
Bennett, C.F., Swayze, E.E., 2010. RNA targeting therapeutics: molecular mechanisms of antisense oligonucleotides as a therapeutic platform. Annu. Rev. Pharmacol. Toxicol. 50, 259–293. https://doi.org/10.1146/annurev.pharmtox.010909.105654.
Boettcher, M., McManus, M.T., 2015. Choosing the right tool for the job: RNAi, TALEN, or CRISPR. Mol. Cell 58, 575–585. https://doi.org/10.1016/j.molcel.2015.04.028.

Bortesi, L., Fischer, R., 2015. The CRISPR/Cas9 system for plant genome editing and beyond. Biotechnol. Adv. 33 (1), 41–52. https://doi.org/10.1016/j.biotechadv.2014.12.006.

Bové, J.M., 2006. Huanglongbing: a destructive, newly-emerging, century-old disease of citrus. J. Plant Pathol. 88, 7–37.

Brookes, G., Barfoot, P., 2016. Global income and production impacts of using GM crop technology 1996–2014. GM Crops Food 7, 38–77. https://doi.org/10.1080/21645698.2016.1176817.

Burand, J.P., Hunter, W.B., 2013. RNAi: future in insect management. J. Invertebr. Pathol. 112, S68–S74. https://doi.org/10.1016/j.jip.2012.07.012.

Capoor, S.P., Rao, G.G., Viswanath, S.M., 1974. In: Weathers, L.G., Cohen, M. (Eds.), Greening disease of citrus in the Deccan trap country and its relationship with the vector, *Diaphorina citri* Kuwayama. Proceedings of 6th Conference of International Organization of Citrus Virologists. University of California, Riverside, CA, pp. 43–49.

Champer, J., Buchman, A., Akbari, O.S., 2016. Cheating evolution: engineering gene drives to manipulate the fate of wild populations. Nat. Rev. Genet. 17, 146–159. https://doi.org/10.1038/nrg.2015.34.

Chaverra-Rodriguez, D., Macias, V.M., Hughes, G.L., Pujhari, S., Suzuki, Y., Peterson, D.R., Kim, D., McKeand, S., Rasgon, J.L., 2018. Targeted delivery of CRISPR-Cas9 ribonucleoprotein into arthropod ovaries for heritable germline gene editing. Nat. Commun. 9, 3008. https://doi.org/10.1038/s41467-018-05425-9.

Chen, L., Wang, G., Zhu, Y.N., Xiang, H., Wang, W., 2016. Advances and perspectives in the application of CRISPR/Cas9 in insects. Zool. Res. 37, 220–228. https://doi.org/10.13918/j.issn.2095-8137.2016.4.220.

Chen, S., Hou, C., Bi, H., Wang, Y., Xu, J., Li, M., et al., 2017. Transgenic CRISPR/Cas9-mediated viral gene targeting for antiviral therapy of *Bombyx mori* nucleopolyhedrovirus (BmNPV). J. Virol. 91, e02465–e02516. https://doi.org/10.1128/JVI.02465-16.

Choudhary, E., Thakur, P., Pareek, M., Agarwal, N., 2015. Gene silencing by CRISPR interference in mycobacteria. Nat. Commun. 6, 6267. https://doi.org/10.1038/ncomms7267.

Chu, C.-C., Gill, T.A., Hoffmann, M., Pelz-Stelinski, K.S., 2016. Inter-population variability of endosymbiont densities in the Asian citrus psyllid (*Diaphorina citri* Kuwayama). Microb. Ecol., 71, 1–9. https://doi.org/10.1007/s00248-016-0733-9.

Cicero, J.M., Alba-Tercedor, J., Hunter, W.B., Saha, S., Brown, S.J., 2018. An animated correspondence of Asian citrus psyllid stylets to the model for biogenesis of potato psyllid stylets. In: Developing an Infrastructure and Product Test Pipeline to Deliver Novel Therapies for Citrus Greening Disease. https://citrusgreening.org/microtomography/cicero_stylet_2018.

Clarke, S.V., Paris, T.M., Hunter, W.B., Brown, S.E., Qureshi, J.A., 2018. Factors affecting citrus tree absorption and RNAi delivery to Asian citrus psyllid. J. Am. Soc. Hortic. Sci. 143 (4). (accepted July12 in press).

Crooke, S.T., 2017. Molecular mechanisms of antisense oligonucleotides. Nucleic Acid Ther. 27 (2), 70–77. https://doi.org/10.1089/nat.2016.0656.

Cui, Y.B., Sun, J.L., Yu, L.L., 2017. Application of the CRISPR gene-editing technique in insect functional genome studies- a review. Entomol. Exp. Appl. 162, 124–132. https://doi.org/10.1111/eea.12530.

da Graça, J.V., Douhan, G.W., Halbert, S.E., Keremane, M.L., Lee, R.F., et al., 2016. Huanglongbing: an overview of a complex pathosystem ravaging the world's citrus. J. Integr. Plant Biol. 58 (4), 373–387.

Daly, S.M., Sturge, C.R., Greenberg, D.E., 2017. Inhibition of bacterial growth by peptide-conjugated morpholino oligomers. In: Moulton, H., Moulton, J. (Eds.), Morpholino Oligomers. In: Methods in Molecular Biology, vol. 1565. Humana Press, New York, NY. https://doi.org/10.1007/978-1-4939-6817-6_10.

Darrington, M., Dalmay, T., Morrison, N.I., Chapman, T., 2017. Implementing the sterile insect technique with RNA interference—a review. Entomol. Exp. Appl. 164, 155–175. https://doi.org/10.1111/eea.12575.

Davey, M.R., Anthony, P., Power, J.B., Lowe, K.C., 2005. Plant protoplasts: status and biotechnological perspectives. Biotechnol. Adv. 23 (2), 131–171.

Dominguez, A.A., Lim, W.A., Qi, L.S., 2016. Beyond editing: repurposing CRISPR-Cas9 for precision genome regulation and interrogation. Nat. Rev. Mol. Cell Biol. 17, 5–15. https://doi.org/10.1038/nrm.2015.2.

Dossi, F.C., da Silva, E.P., Consoli, F.L., 2014. Population dynamics and growth rates of endosymbionts during *Diaphorina citri* (Hemiptera, Liviidae) ontogeny. Microb. Ecol. 68, 881–889. https://doi.org/10.1007/s00248-014-0463-9.

Doudna, J.A., Charpentier, E., 2014. The new frontier of genome engineering with CRISPR-Cas9. Science 346 (6213), 1258096. https://doi.org/10.1126/science.1258096.

Dowling, D., Pauli, T., Donath, A., Meusemann, K., Podsiadlowski, L., Petersen, M., Peters, R.S., Mayer, C., Liu, S., Zhou, X., Misof, B., Niehuis, O., 2016. Phylogenetic origin and diversification of RNAi pathway genes in insects. Genome Biol. Evol. 8 (12), 3784–3793. https://doi.org/10.1093/gbe/evw281.

Dubelman, S., Fischer, J., Zapata, F., Huizinga, K., Jiang, C., Uffman, J., et al., 2014. Environmental fate of double-stranded RNA in agricultural soils. PLoS One 9, 1–7.

Dutt, M., Barthe, G., Irey, M., Grosser, J., Duan, Y., et al., 2015. Transgenic citrus expressing an *Arabidopsis* NPR1 gene exhibit enhanced resistance against Huanglongbing (HLB; citrus greening). PLoS One 10 (9), e0137134.

Eguchi, A., Meade, B.R., Chang, Y.-C., Fredrickson, C.T., Willert, K., Puri, N., et al., 2009. Efficient siRNA delivery into primary cells by a peptide transduction domain-dsRNA binding domain fusion protein. Nat. Biotechnol. 27, 567–571. https://doi.org/10.1038/nbt.1541.

Elbashir, S.M., Lendeckel, W., Tuschl, T., 2001. RNA interference is mediated by 21- and 22-nucleotide RNAs. Genes Dev. 15, 188–200.

Eleftherianos, I., Atri, J., Accetta, J., Castillo, J.C., 2013. Endosymbiotic bacteria in insects: guardians of the immune system? Front. Physiol. 4, 46. https://doi.org/10.3389/fphys.2013.00046.

El-Shesheny, I., Hajeri, S., El-Hawary, I., Gowda, S., Killiny, N., 2013. Silencing abnormal wing disc gene of the Asian citrus psyllid, *Diaphorina citri* disrupts adult wing development and increases nymph mortality. PLoS One 8, e65392.

Faustinelli, P.C., Power, I.L., Arias, R.S., 2018. Detection of exogenous double-stranded RNA movement in in vitro peanut plants. Plant Biol. (Stuttg) 20, 444–449. https://doi.org/10.1111/plb.12703.

Fire, A., Xu, S., Montgomery, M.K., Kostas, S.A., Driver, S.E., Mello, C.C., 1998. Potent and specific genetic interference by double-stranded RNA in *Caenorhabditis elegans*. Nature 391, 806–811.

Forster, A.C., Altman, S., 1990. External guide sequences for an RNA enzyme. Science 249, 783–786.

Gaj, T., Gersbach, C.A., Barbas III, C.F., 2013. ZFN, TALEN, and CRISPR/Cas-based methods for genome engineering. Trends Biotechnol, 31, 397–405. https://doi.org/10.1016/j.tibtech.2013.04.004.

Galdeano, D.M., Breton, M.C., Lopes, J.R.S., Falk, B.W., Machado, M.A., 2017. Oral delivery of double-stranded RNAs induces mortality in nymphs and adults of the Asian citrus psyllid, *Diaphorina citri*. PLoS One 12 (3), e0171847. https://doi.org/10.1371/journal.pone.0171847.

Gantz, V.M., Akbari, O.S., 2018. Gene editing technologies and applications for insects. Curr. Opin. Insect Sci. 28, 66–72. https://doi.org/10.1016/j.cois.2018.05.006.

Garczynski, S.F., Martin, J.A., Griset, M., Willett, L.S., Cooper, W.R., Swisher, K.D., Unruh, T.R., 2017. CRISPR/Cas9 editing of the codling moth (Lepidoptera: Tortricidae) CpomOR1 gene affects egg production and viability. J. Econ. Entomol. 110 (4), 1847–1855. https://doi.org/10.1093/jee/tox166.

Garneau, J.E., Dupuis, M., Viooion, M., Romero, D., Barrangou, R., Boyaval, P., Fremaux, C., Horvath, P., Magadan, A., Moineau, S., 2010. The CRISPR/Cas bacterial immune system cleaves bacteriophage and plasmid DNA. Nature 468, 67–71.

Ghanim, M., Fattah-Hosseini, S., Levy, A., Cilia, M., 2016. Morphological abnormalities and cell death in the Asian citrus psyllid (*Diaphorina citri*) midgut associated with Candidatus Liberibacter asiaticus. Sci. Rep. 6, 33418. https://doi.org/10.1038/srep33418. PMID: 27630042.

Ghosh, S.K.B., Hunter, W.B., Park, A.L., Gundersen-Rindal, D.E., 2017. Double strand RNA delivery system for plant-sap-feeding insects. PLoS One 12 (2), e0171861. https://doi.org/10.1371/journal.pone.0171861.

Ghosh, S.K., Hunter, W.B., Park, A.L., Gundersen-Rindal, D.E., 2018. Double-stranded RNA oral delivery methods to induce RNA interference in phloem and plant-sap-feeding hemipteran insects. J. Vis. Exp. (135), e57390. https://doi.org/10.3791/57390.

Gillet, F.-X., Garcia, R.A., Macedo, L.L.P., Albuquerque, E.V.S., Silva, M.C.M., Grossi-de-Sa, M.F., 2017. Investigating engineered ribonucleoprotein particles to improve oral RNAi delivery in crop insect pests. Front. Physiol. 8, 256. https://doi.org/10.3389/fphys.2017.00256.

Gottwald, T.R., 2010. Current epidemiological understanding of citrus huanglongbing. Annu. Rev. Phytopathol. 48, 119–139. https://doi.org/10.1146/annurev-phyto-073009-114418.

Grafton-Cardwell, E.-E., Stelinski, L.L., Stansly, P.A., 2013. Biology and management of Asian citrus psyllid, vector of the huanglongbing pathogens. Annu. Rev. Entomol. 58, 413–432.

Gupta, S.K., Shukla, P., 2016. Gene editing for cell engineering: trends and applications. Crit. Rev. Biotechnol. 37 (5), 672–684. https://doi.org/10.1080/07388551.2016.1214557.

Haapalainen, M., 2014. Biology and epidemics of Candidatus Liberibacter species, psyllid-transmitted plant-pathogenic bacteria. Ann. Appl. Biol. 165 (2), 172–198.

Hagedorn, P.H., Pontoppidan, M., Bisgaard, T.S., Berrera, M., Dieckmann, A., et al., 2018. Identifying and avoiding off-target effects of RNase H-dependent antisense oligonucleotides in mice. Nucleic Acids Res. 46 (11), 5366–5380. https://doi.org/10.1093/nar/gky397.

Hajeri, S., Killiny, N., El-Mohtar, C., Dawson, W.O., Gowda, S., 2014. Citrus tristeza virus-based RNAi in citrus plants induces gene silencing in *Diaphorina citri*, a phloem-sap sucking insect vector of citrus greening disease (Huanglongbing). J. Biotechnol. 176, 42–49. Elsevier B.V.

Halbert, S.E., Manjunath, K.L., 2006. Asian citrus psyllids (Sternorrhyncha: Psyllidae) and greening disease of citrus: a literature review and assessment of risk in Florida. Fla. Entomol. 87, 330–353.

Hall, D.G., Richardson, M.L., Ammar, E.-D., Halbert, S.E., 2013. Asian citrus psyllid, *Diaphorina citri*, vector of citrus huanglongbing disease. Entomol. Exp. Appl. 146, 207–223.

Hannon, G.J., 2002. RNA interference. Nature 418, 244–251.

Hansen, A.K., Trumble, J.T., Stouthamer, R., Paine, T.D., 2008. A new huanglongbing species, 'Candidatus Liberibacter psyllaurous' found to infect tomato and potato, is vectored by the psyllid *Bactericera cockerelli* (Sulc). Appl. Environ. Microbiol. 74, 5862–5865.

Hao, G., Stover, E., Gupta, G., 2016. Overexpression of a modified plant thionin enhances disease resistance to citrus canker and Huanglongbing (HLB). Front. Plant Sci. 7, 1078.

Hegarty, J.P., Stewart, E.B., 2018. Advances in therapeutic bacterial antisense biotechnology. Appl. Microbiol. Biotechnol. 102, 1055–1065. https://doi.org/10.1007/s00253-017-8671-0.

Hoddle, M.S., Pandey, R., 2014. Host range testing of *Tamarixia radiata* (Hymenoptera: Eulophidae) sourced from the Punjab of Pakistan for classical biological control of *Diaphorina citri* (Hemiptera: Liviidae: Euphyllurinae: Diaphorinini) in California. J. Econ. Entomol. 107, 125–136. https://doi.org/10.1603/EC13318. PMID: 24665694.

Hudziak, R.M., Barofsky, E., Barofsky, D.F., Weller, D.L., Huang, S.B., Weller, D.D., 1996. Resistance of Morpholino phosphorodiamidate oligomers to enzymatic degradation. Antisense Nucleic Acid Drug Dev. 6, 267–272.

Hunter, W.B., Dowd, S.E., Katsar, C.S., Shatters, R.G., McKenzie, C.L., Hall, D.G., 2009. Psyllid biology: expressed genes in adult Asian citrus psyllids, *Diaphorina citri* Kuwayama. Open Entomol. J. 3, 18–29.

Hunter, W.B., Avery, P.B., Pick, D., Powell, C.A., 2011. Broad spectrum potential of *Isaria fumosorosea* on insect pests of citrus. Fla. Entomol. 94, 1051–1054.

Hunter, W.B., Glick, E., Paldi, N., Bextine, B.R., 2012. Advances in RNA interference: dsRNA treatment in trees and grapevines for insect pest population suppression. Southwest. Entomol. 37 (1), 85–87. https://doi.org/10.3958/059.037.0110.

Hunter, W.B., Gonzalez, M.T., Andrade, E.C. 2017. Double stranded RNA compositions for reducing Asian citrus Psyllid infestation and methods of use. U.S. Patent Application Publication US 2017/0211082 Al.

Hunter, W.B., Lopez, S.L., 2018. Improved RNAi persistence and activity by incorporating non-canonical nucleotides into dsRNA. Biological Archives. bioRxiv. (in prep).

Hunter, W.B., et al., 2018a. Biotechnology: RNAi, antisense oligonucleotides and CRISPR strategies to reduce psyllids and bacterial pathogens in citrus. In: Reibach, P., Ruebelt, M. (Eds.), AGRO, RNAi & Gene Editing: Utilization for enhanced Crop Production. Proceedings of the 256th American Chemical Society, 19–23 August, Boston, MA.

Hunter, W.B., Gonzalez, M.T., Tomich, J., 2018b. BAPC–CRISPR/Cas9 System Injects Ovaries for Heritable Germline Gene Editing (Asian Citrus Psyllid: Liviidae: Hemiptera). (in prep).

Hunter, W.B., Holland, C.S., Lopez, S.L., Boyle, M.J., 2018c. Method for quantification and location of fluorescence G-labeled dsRNA in specific tissues of Asian citrus psyllid and citrus trees. Biological Archives. bioRxiv. (in prep).

Huvenne, H., Smagghe, G., 2010. Mechanisms of dsRNA uptake in insects and potential of RNAi for pest control: a review. J. Insect Physiol. 56, 227–235.

Inoue, H., Ohnishi, J., Ito, T., Tomimura, K., Miyata, S., Iwanami, T., et al., 2009. Enhanced proliferation and efficient transmission of Candidatus Liberibacter asiaticus by adult *Diaphorina citri* after acquisition feeding in the nymphal stage. Ann. Appl. Biol. 155, 29–36. https://doi.org/10.1111/j.1744-7348.2009.00317.x.

Ivashuta, S.I., Petrick, J.S., Heisel, S.E., Zhang, Y., Guo, L., Reynolds, T.L., Rice, J.F., Allen, E., Roberts, J.K., 2009. Endogenous small RNAs in grain: semi-quantification and sequence homology to human and animal genes. Food Chem. Toxicol. 47, 353–360.

Ivashuta, S., Zhang, Y., Wiggins, B.E., Ramaseshadri, P., Segers, G.C., Johnson, S., Meyer, S.E., Kerstetter, R.A., McNulty, B.C., Bolognesi, R., Heck, G.R., 2015. Environmental RNAi in herbivorous insects. RNA 21, 840–850. https://doi.org/10.1261/rna.048116.114.

James, C., 2015. Global Status of Commercialized Biotech/GM Crops 2015: ISAAA Brief No. 51. ISAAA, Ithaca, NY. http://isaaa.org/resources/publications/briefs/51/executivesummary/default.asp.

Jarrous, N., 2017. Roles of RNase P and its subunits. Trends Genet. 33, 9. https://doi.org/10.1016/j.tig.2017.06.006.

Jarvis, D.L., Oker-Blom, C., Summers, M.D., 1990. Role of glycosylation in the transport of recombinant glycoproteins through the secretory pathway of lepidopteran insect cells. J. Cell. Biochem. 42, 181–191.

Jiang, W., Bikard, D., Cox, D., Zhang, F., Marraffini, L.A. 2013. RNA-guided editing of bacterial genomes using CRISPR-Cas systems. Nat. Biotechnol. 31(3):233–9. https://doi.org/10.1038/nbt.2508. PMC 3748948 Freely accessible. PMID 23360965.

Jinak, M., Doudna, J.A., 2009. A three-dimensional view of the molecular machinery of RNA interference. Nature 457 (7228), 405–412. https://doi.org/10.1038/nature07755.

Joga, M.R., Zotti, M.J., Smagghe, G., Christiaens, O., 2016. RNAi efficiency, systemic properties, and novel delivery methods for pest insect control: what we know so far. Front. Physiol. 7, 553. https://doi.org/10.3389/fphys.2016.00553.

Juliano, R.L., 2016. The delivery of therapeutic oligonucleotides. Nucleic Acids Res. 44, 6518–6548. https://doi.org/10.1093/nar/gkw236. Survey and Summary.

Kanakala, S., Ghanim, M., 2016. RNA interference in insect vectors for plant viruses. Viruses 8, 329. https://doi.org/10.3390/v8120329.

Karl, H., Gordon, J., Waterhouse, P.M., 2007. RNAi for insect-proof plants. Nat. Biotechnol. 25, 1231–1232.

Killiny, N., Kishk, A., 2017. Delivery of dsRNA through topical feeding for RNA interference in the citrus sap piercing-sucking hemipteran, Diaphorina citri. Arch. Insect Biochem. Physiol. 95, e21394.

Killiny, N., Hajeri, S., Tiwari, S., Gowda, S., Stelinski, L.L., 2014. Double-stranded RNA uptake through topical application, mediates silencing of five CYP4 genes and suppresses insecticide resistance in *Diaphorina citri*. PLoS One 9, 1–8.

Kishk, A., Anber, H.A.I., Abdel-Raof, T.K., El-Sherbeni, A.-H.D., Hamed, S., Gowda, S., Killiny, N., 2017. RNA interference of carboxylesterases causes nymph mortality in the Asian citrus psyllid, Diaphorina citri. Arch. Insect Biochem. Physiol. 94, e21377. https://doi.org/10.1002/arch.21377.

Koch, M.S., Ward, J.M., Levine, S.L., Baum, J.A., Vicini, J.L., Hammond, B.G., 2015. The food and environmental safety of Bt crops. Front. Plant Sci. 6, 283. https://doi.org/10.3389/fpls.2015.00283.

Koch, A., Biedenkopf, D., Furch, A., Weber, L., Rossbach, O., Abdellatef, E., et al., 2016. An RNAi-based control of *Fusarium graminearum* infections through spraying of long dsRNAs involves a plant passage and is controlled by the fungal silencing machinery. PLoS Pathog. 12 (10), e1005901. https://doi.org/10.1371/journal.ppat.1005901.

Kolliopoulou, A., Taning, C.N.T., Smagghe, G., Swevers, L., 2017. Viral delivery of dsRNA for control of insect agricultural pests and vectors of human disease: prospects and challenges. Front. Physiol. 8, 399. https://doi.org/10.3389/fphys.2017.00399.

Kruse, A., Fattah-Hosseini, S., Saha, S., Johnson, R., Warwick, E., Sturgeon, K., et al., 2017. Combining'omics and microscopy to visualize interactions between the Asian citrus psyllid vector and the Huanglongbing pathogen Candidatus Liberibacter asiaticus in the insect gut. PLoS One 12 (6), e0179531. https://doi.org/10.1371/journal.pone.0179531.

Kupferschmidt, K., 2013. A lethal dose of RNA. Science 341, 732–733.

Kurreck, J., 2003. Antisense technologies: improvement through novel chemical modifications. Eur. J. Biochem. 270, 1628–1644. https://doi.org/10.1046/j.1432-1033.2003.03555.x. [PubMed] [Cross Ref].

Kyndt, T., Quispe, D., Zhai, H., Jarret, R., et al., 2015. The genome of cultivated sweet potato contains A. tumefaciens T-DNAs with expressed genes: an example of a naturally transgenic food crop. PNAS 112 (8), 5844–5849. https://doi.org/10.1073/pnas.1419685112.

Lacey, L.A., Grxywacz, D., Shapiro-Ilan, D.I., Frutos, R., Brownbridge, M., Goettel, M.S., 2015. Insect pathogens as biological control agents: back to the future. J. Invertebr. Pathol. 132, 1–41. https://doi.org/10.1016/j.jip.2015.07.009.

Lallemand, J., Fos, A., Bové, J.M., 1986. Transmission de la bactérie associée à la forme africaine de le maladie du "Greening" par le psylla asiatique *Diaphorina citri* Kuwayama. Fruits 42, 341–343.
Larson, M.H., Gilbert, L.A., Wang, X., Lim, W.A., Weissman, J.S., Qi, L.S., 2013. CRISPR interference (CRISPRi) for sequence-specific control of gene expression. Nat. Protoc. 8 (11), 2180–2196. https://doi.org/10.1038/nprot.2013.132.
La Russa, M.F., Qi, L.S., 2015. The new state of the art: Cas9 for gene activation and repression. Mol. Cell. Biol. 35, 3800–3809. https://doi.org/10.1128/MCB.00512-15.
Li, J., Wang, X.P., Wang, M.Q., Ma, W.H., Hua, H.X., 2013. Advances in the use of the RNA interference technique in Hemiptera. Insect Sci. 20, 31–39. https://doi.org/10.1111/j.1744-7917.2012.01550.x.
Li, H., Guan, R., Guo, H., Miao, X., 2015. New insights into an RNAi approach for plant defence against piercing-sucking and stem-borer insect pests. Plant Cell Environ. 38 (11), 2277–2285. https://doi.org/10.1111/pce.12546.
Li, X., Jun, Y., Erickstad, M., Brown, S., Parks, A., Court, D; Jun, S. 2016. tCRISPRi: tunable and reversible, one-step control of gene expression. Sci. Rep. 6:39096. Doi.https://doi.org/10.1038/srep39076. PMC 5171832.
Liang, X., Potter, J., Kumar, S., Zou, Y., Quintanilla, R., et al., 2015. Rapid and highly efficient mammalian cell engineering via Cas9 protein transfection. J. Biotechnol. 208, 44–53. https://doi.org/10.1016/j.jbiotec.2015.04.024.
Lombardo, L., Coppola, G., Zelasco, S., 2016. New technologies for insect-resistant and herbicide-tolerant plants. Trends Biotechnol. 34, 49–57. https://doi.org/10.1016/j.tibtech.2015.10.006.
Ludin, K.S., Gissberg, O., Edvard Smith, C.I., 2015. Oligonucleotide therapies: the past and the present. Hum. Gene Ther. 26 (8), 475–485. https://doi.org/10.1089/hum.2015.070.
Markert, M.J., Zhang, Y., Enuameh, M.S., Reppert, S.M., Wolfe, S.A., Merlin, C., 2016. Genomic access to monarch migration using TALEN and CRISPR/Cas9-mediated targeted mutagenesis. G3 (Bethesda) 6, 905–915.
Marshall, J.M., Akbari, O.S., 2016. Gene drive strategies for population replacement. Genet. Control Malaria Dengue 2016, 169–200.
Marshall, J.M., Akbari, O.S., 2018. Can CRISPR-based gene drive be confined in the wild? A question for molecular and population biology. ACS Chem. Biol. 13, 424–430. https://doi.org/10.1021/acschembio.7b00923.
Marutani-Hert, M., Hunter, W.B., Hall, D.G. 2008. Novel reovirus in *Diaphorina citri* (Hemiptera: Psyllidae). 11.5, pg. 320–323. IRCHLB Proc. Dec. www.plantmanagementnetwork.org.
Marutani-Hert, M., Hunter, W.B., Katsar, C.S., Sinisterra, X.H., Hall, D.G., Powell, C.A., 2009. Reovirus-like sequences isolated from adult Asian citrus psyllid, (Hemiptera: Psyllidae: *Diaphorina citri*). Fla. Entomol. 92, 314–320.
Massonié, G., Garnier, M., Bové, J.M., 1976. In: Transmission of Indian citrus decline by *Trioza erytreae* (Del Guercio), the vector of South African reening. Proc. Conf. Int. Organ. Citrus Virol. 7th, Athens. IOCV, Riverside, CA, pp. 18–20.
McGhee, G.C., Guasco, J., Bellomo, L.M., Blumer-Schuette, S.E., Shane, W.W., Irish-Brown, A., et al., 2011. Genetic analysis of streptomycin-resistant (Sm(R)) strains of *Erwinia amylovora* suggests that dissemination of two genotypes is responsible for the current distribution of Sm(R) *E. amylovora* in Michigan. Phytopathology 101, 182–191. https://doi.org/10.1094/PHYTO-04-10-0127.
Meyer, J.M., Hoy, M.A., 2008. Molecular survey of endosymbionts in Florida populations of *Diaphorina citri* (Hemiptera: Psyllidae) and its parasitoids *Tamarixia radiata* (Hymenoptera: Eulophidae) and *Diaphorencyrtus aligarhensis* (Hymenoptera: Encyrtidae). Fla. Entomol. 91, 294–304.
Michaud, J.P., Olsen, L.E., 2004. Suitability of Asian citrus psyllid, *Diaphorina citri*, as prey for ladybeetles. BioControl 49, 417–431.

Mitter, N., Worrall, E.A., Robinson, K.E., Li, P., Jain, R.G., Taochy, C., Fletcher, S.J., Carroll, B.J., Lu, G.Q., Xu, Z.P., 2017. Clay nanosheets for topical delivery of RNAi for sustained protection against plant viruses. Nat. Plants 3 (16207). https://doi.org/10.1038/nplants.2016.207.
Moulton, J.D., 2016. Guide for morpholino users: toward therapeutics. J. Drug Discov. Dev. Deliv. 3 (2), 1023. ISSN: 2471-0288.
Moulton, H., Moulton, J., 2017. Morpholino oligomers. In: Moulton, H., Moulton, J. (Eds.), In: Methods in Molecular Biology, vol. 1565. Humana Press, New York, NY.
Moulton, J.D., Yan, Y.-L., 2008. Using morpholinos to control gene expression, Unit 26.8. Curr. Protoc. Mol. Biol. 26.8.1–26.8.29. https://doi.org/10.1002/0471142727.mb2608s83.
Moulton, H.M., Nelson, M.H., Hatlevig, S.A., Reddy, M.T., Iversen, P.L., 2004. Cellular uptake of antisense Morpholino oligomers conjugated to arginine-rich peptides. Bioconjug. Chem. 15, 290–299.
Mout, R., Tonga, G.Y., Wang, L.-S., Ray, M., Roy, T., Rotello, W.M., 2017. Programmed self-assembly of hierarchical nanostructures through protein – nanoparticle coengineering. ACS Nano 11, 3456–3462. https://doi.org/10.1021/acsnano.6b07258.
Munyaneza, J.E., Sengoda, V.G., Buchman, J.L., Fisher, T.W., 2012. Effects of temperature on 'Candidatus Liberibacter solanacearum' and zebra chip potato disease symptom development. Plant Dis. 96, 18–23.
Murphy, K.A., Tabuloc, C.A., Cervantes, K.R., Chiu, J.C., 2016. Ingestion of genetically modified yeast symbiont reduces fitness of an insect pest via RNA interference. Sci. Rep. 6, 22587. https://doi.org/10.1038/srep22587.
Mysore, K., Andrews, E., Li, P., Duman-Scheel, M., 2014. Chitosan/siRNA nanoparticle targeting demonstrates a requirement for single-minded during larval and pupal olfactory system development of the vector mosquito *Aedes aegypti*. BMC Dev. Biol. 14, 1. https://doi.org/10.1186/1471-213X-14-9.
Nakabachi, A., Yamashita, A., Toh, H., Ishikawa, H., Dunbar, H.E., Moran, N.A., Hattori, M., 2006. The 160-kilobase genome of the bacterial endosymbiont Carsonella. Science 314, 267.
Nakabachi, A., Ueoka, R., Oshima, K., Teta, R., Mangoni, A., Gurgui, M., Oldham, N.J., van Echten-Deckert, G., Okamura, K., Yamamoto, K., Inoue, H., Ohkuma, M., Hongoh, Y., Miyagishima, S.-Y., Hattori, M., Piel, J., Fukatsu, T., 2013. Defensive bacteriome symbiont with a drastically reduced genome. Curr. Biol. 23, 1478–1484.
Napoli, C., Lemieux, C., Jorgensen, R., 1990. Introduction of a chimeric chalcone synthase gene into petunia results in reversible co-suppression of homologous genes in trans. Plant Cell 2, 279–289.
NASEM, National Academies of Sciences, Engineering, and Medicine, 2016. Genetically Engineered Crops: Experiences and Prospects. The National Academies Press, Washington, DC.https://doi.org/10.17226/23395 Available:http://nap.edu/23395.
Norelli, J.L., Jones, A.L., Aldwinckle, H.S., 2003. Fire blight management in the twenty-first century: using new technologies that enhance host resistance in apple. Plant Dis. 87, 756–765. https://doi.org/10.1094/PDIS.2003.87.7.756.
Nouri, S., Salem, N., Nigg, J.C., FalK, B.W., 2015. A diverse array of new viral sequences identified in worldwide populations of the Asian citrus psyllid (*Diaphorina citri*) using viral metagenomics. J. Virol. 90, 2434–2445. https://doi.org/10.1128/JVI.02793-15.
Pan, X., Zhou, G., Wu, J., Bian, G., Lu, P., Raikhel, A.S., Xi, Z., 2012. *Wolbachia* induces reactive oxygen species (ROS)-dependent activation of the Toll pathway to control dengue virus in the mosquito *Aedes aegypti*. Proc. Natl. Acad. Sci. U. S. A. 109, E23–E31.
Patel, R.R., Sundin, G.W., Yang, C.-H., Wang, J., Huntley, R.B., Yuan, X., Zeng, Q., 2017. Exploration of using antisense peptide nucleic acid (PNA)-cell penetrating peptide (CPP) as a novel bactericide against fire blight pathogen *Erwinia amylovora*. Front. Microbiol. 8, 687. https://doi.org/10.3389/fmicb.2017.00687.

Patt, J.M., Chow, A., Meikle, W.G., Gracia, C., Jackson, M.A., Flores, D., Sétamou, M., Dunlap, C.A., Avery, P.B., Hunter, W.B., Adamczyk, J.J., 2015. Efficacy of an autodisseminator of an entomopathogenic fungus, *Isaria fumosorosea*, to suppress Asian citrus psyllid, *Diaphorina citri*, under greenhouse conditions. Biol. Control. 88, 37–45. https://doi.org/10.1016/j.biocontrol.2015.04.014.

Pelz-Stelinski, K.S., Killiny, N., 2016. Better together: association with "Candidatus Liberibacter asiaticus" increases the reproductive fitness of its insect vector, *Diaphorina citri* (Hemiptera: Liviidae). Ann. Entomol. Soc. Am. 109, 371–376.

Pelz-Stelinski, K.S., Brlansky, R.H., Ebert, T.A., Rogers, M.E., 2010. Transmission parameters for Candidatus Liberibacter asiaticus by Asian citrus psyllid (Hemiptera: Psyllidae). J. Econ. Entomol. 103, 1531–1541.

Pelz-Stelinski, K.S., Hunter, W.B., Altman, S., 2017. Targeting Microbes to Control Huanglongbing Disease of Citrus. USDA, NIFA. Review.

Peters, J.M., Colavin, A., Shi, H., et al. (14). 2016. A comprehensive, CRISPR-based functional analysis of essential genes in bacteria. Cell 165 (6):1493–1506. https://doi.org/10.1016/j.cell.2016.05.003. PMC 4894308 Freely accessible. PMID 27238023.

Petrick, J.S., Brower-Toland, B., Jackson, A.L., Kier, L.D., 2013. Safety assessment of food and feed from biotechnology-derived crops employing RNA-mediated gene regulation to achieve desired traits: a scientific review. Regul. Toxicol. Pharmacol. 66, 167–176. https://doi.org/10.1016/j.yrtph.2013.03.008.

Petrick, J.S., Frierdich, G.E., Carleton, S.M., Kessenich, C.R., Silvanovich, A., Zhang, Y., Koch, M.S., 2016. Corn rootworm-active RNA DvSnf7: repeat dose oral toxicology assessment in support of human and mammalian safety. Regul. Toxicol. Pharmacol. 81, 57e68. https://doi.org/10.1016/j.yrtph.2016.07.009.

Price, D.R., Gatehouse, J.A., 2008. RNAi-mediated crop protection against insects. Trends Biotechnol. 26, 393–400. https://doi.org/10.1016/j.tibtech.2008.04.004.

Qi, L.S., Larson, M.H., Gilbert, L.A., Doudna, J.A., Weissman, J.S., Arkin, A.P., Lim, W.A. 2013. Repurposing CRISPR as an RNA-guided platform for sequence-specific control of gene expression. Cell 152 (5):1173–83. https://doi.org/10.1016/j.cell.2013.02.022. PMC 3664290 Freely accessible. PMID 23452860.

Raiol-Junior, L.L., Baia, A.D.B., Luiz, F.Q.B.F., Fassini, C.G., Marques, V.V., Lopes, S.A., 2017. Improvement in the excised citrus leaf assay to investigate inoculation of 'Candidatus Liberibacter asiaticus' by the Asian citrus psyllid *Diaphorina citri*. Plant Dis. 101, 409–413. https://doi.org/10.1094/PDIS-08-16-1093-RE.

Ramsey, J.S., Johnson, R.S., Hoki, J.S., Kruse, A., Mahoney, J., Hilf, M.E., Hunter, W.B., Hall, D.G., Schroeder, F.C., MacCoss, M.J., Cilia, M., 2015. Metabolic interplay between the Asian citrus psyllid and its Profftella symbiont: an Achilles' heel of the citrus greening insect vector. PLoS One 10, e0140826–e0140921.

Rasmussen, L.C., Sperling-Petersen, H.U., Mortensen, K.K., 2007. Hitting bacteria at the heart of the central dogma: sequence-specific inhibition. Microb. Cell Fact. 6, 24. https://doi.org/10.1186/1475-2859-6-24.

Ren, S.-L., Li, Y.-H., Zhou, Y.-T., Xu, W.-M., Cuthbertson, A.G., Guo, Y.-J., Qiu, B.-L., 2016. Effects of Candidatus Liberibacter asiaticus on the fitness of the vector Diaphorina citri. J. Appl. Microbiol. 121, 1718–1726. https://doi.org/10.1111/jam.13302.

Ricroch, A.E., Hénard-Damave, M.C., 2016. Next biotech plants: new traits, crops, developers and technologies for addressing global challenges. Crit. Rev. Biotechnol. 36 (4), 675–690. https://doi.org/10.3109/07388551.2015.1004521.

Roberts, A.F., Devos, Y., Lemgo, G.N.Y., Zhou, X., 2015. Biosafety research for non-target organism risk assessment of RNAi-based GE plants. Front. Plant Sci. 6, 958. https://doi.org/10.3389/fpls.2015.00958.

Rodrigues, T.B., Figueira, A., 2016. Management of insect pest by RNAi—a new tool for crop protection. In: Abdurakhmonov, I.Y. (Ed.), RNA Interference. InTech. https://doi.org/10.5772/61807.

Russell, C.W., Pelz-Stelinski, K.S., 2015. Development of an artificial diet and feeding system for juvenile stages of the Asian citrus psyllid, *Diaphorina citri*. Entomol. Exp. Appl. 154, 171–176.

Saha, S., Hunter, W.B., Reese, J., Morgan, J.K., Marutani-Hert, M., Huang, H., Lindeberg, M., 2012. Survey of endosymbionts in the *Diaphorina citri* metagenome and assembly of a *Wolbachia* wDi draft genome. PLoS One 7 (11), e50067.

Saha, S., Hosmani, P.S., Villalobos-Ayala, K., Miller, S., et al., 2017a. Improved annotation of the insect vector of Citrus greening disease: biocuration by a diverse genomics community. bioRxiv. https://doi.org/10.1101/099168.

Saha, S., Hosmani, P.S., Flores-Gonzalez, M., Hunter, W., D'Elia, T. 2017b. Biocuration and improvement of the *Diaphorina citri* draft genome assembly with long reads, optical maps and long-range scaffolding. No. 13. Pg. 27–28. Proc. 10th Arthropod Genomics Symposium, Jun. 8–11, Notre Dame, IN. http://globalhealth.nd.edu/assets/238736/ags_booklet.pdf.

Sala, C.D., Soler-Bistué, A.J.C., Korprapun, L., Zorreguieta, A., Tolmasky, M.E., 2012. Inhibition of cell division induced by external guide sequences (EGS Technology) targeting ftsZ. PLoS One 7 (10), e47690. https://doi.org/10.1371/journal.pone.0047690.

San Miguel, K., Scott, J.G., 2016. The next generation of insecticides: DsRNA is stable as a foliar-applied insecticide. Pest Manag. Sci. 72, 801–809.

Sandoval-Mojica, A.F., Hunter, W.B., Pelz-Stelinski, K.S., 2017. In: Targeting Asian citrus psyllid microbiota for management of Huanglongbing. P-IE Section Symposium: Understanding the Host-Vector-Pathogen Interactions Involved in Citrus Greening Disease: Advances in Vector and Disease Control. Entomological Soc. America Annual Conf., Denver, CO, Nov. 5–8.

Saurabh, S., Vidyarthi, A.S., Prasad, D., 2014. RNA interference: concept to reality in crop improvement. Planta 239, 543–564. https://doi.org/10.1007/s00425-013-2019-5.

Scott, J.G., Michel, K., Bartholomay, L.C., Siegfried, B.D., Hunter, W.B., Smagghe, G., Zhu, K.Y., Douglas, A.E., 2013. Towards the elements of successful insect RNAi. J. Insect Physiol. 59, 1212–1221. Elsevier Ltd.

Sen, G.L., Blau, H.M., 2006. A brief history of RNAi: the silence of the genes. FASEB J. 20, 1293–1299.

Shen, X., Corey, D.R., 2018. Chemistry, mechanism and clinical status of antisense oligonucleotides and duplex RNAs. Nucleic Acids Res. 46 (4), 1584–1600. https://doi.org/10.1093/nar/gkx1239. Survey and Summary.

Sherman, J.H., Munyikwa, T., Chan, S.Y., Petrick, J.S., Witwer, K.W., Choudhuri, S., 2015. RNAi technologies in agricultural biotechnology: the Toxicology Forum 40th Annual Summer Meeting. Regul. Toxicol. Pharmacol. 73, 671–680. https://doi.org/10.1016/j.yrtph.2015.09.001.

Singh, I.K., Singh, S., Mogilicherla, K., Shukla, J.N., Palli, S.R., 2017. Comparative analysis of double-stranded RNA degradation and processing in insects. Sci. Rep. 7, 17059. https://doi.org/10.1038/s41598-017-17134-2.

Sinisterra-Hunter, X., Hunter, W.B. 2018. Towards a holistic integrated pest management: lessons learned from plant-insect mechanisms in the field. Pp. 204-226. In The Biology of Plant-Insect Interactions: A Compendium for the Plant Biotechnologist. 236 Pgs. Chandrakanth Emani (ed). ISBN 9781498709736-CAT#K25008 (Chapter 10). CRC Press.

Stover, E., Stange, R.R., McCollum, T.G., Jaynes, J., Irey, M., Mirkov, E., 2013. Screening antimicrobial peptides in vitro for use in developing transgenic citrus resistant to huanglongbing and citrus canker. J. Am. Soc. Hortic. Sci. 138 (2), 142–148.

Subandiyah, S., Nikoh, N., Tsuyumu, S., Somowiyarjo, S., Fukatsu, T., 2000. Complex endosymbiotic microbiota of the citrus psyllid Diaphorina citri (Homoptera: Psylloidea). Zool. Sci. 17, 983–989. https://doi.org/10.2108/zsj.17.983.

Sukthankar, P., Avila, L.A., Whitaker, S.K., Iwamoto, T., Morgenstern, A., Apostolidis, C., Liu, K., Hanzlik, R.P., Dadachova, E., Tomich, J.M., 2014. Branched amphiphilic peptide capsules: cellular uptake and retention of encapsulated solutes. Biochim. Biophys. Acta 1838, 2296–2305. https://doi.org/10.1016/j.bbamem.2014.02.005.

Summerton, J., Weller, D., 1997. Morpholino antisense oligomers: design, preparation, and properties. Antisense Nucleic Acid Drug Dev. 7, 187–195.

Sun, D., Guo, Z., Liu, Y., Zhang, Y., 2017. Progress and prospects of CRISPR/Cas systems in insects and other arthropods. Front. Physiol. 8, 608. https://doi.org/10.3389/fphys.2017.00608.

Taning, C.N.T., Andrade, E.C., Hunter, W.B., Christiaens, O., Smagghe, G., 2016. Asian citrus psyllid RNAi pathway—RNAi evidence. Sci. Rep. 6, 38082. https://doi.org/10.1038/srep38082.

Taning, C.N.T., van Eynde, B., Yu, N., Ma, S., Smagghe, G., 2017. CRISPR/Cas9 in insects: applications, best practices and biosafety concerns. J. Insect Physiol. 98, 245–257. https://doi.org/10.1016/j.jinsphys.2017.01.007.

Terenius, O., Papanicolaou, A., Garbutt, J.S., Eleftherianos, I., Huvenne, H., Kanginakudru, S., Albrechtsen, M., An, C., Aymeric, J.L., Barthel, A., et al., 2011. RNA interference in Lepidoptera: an overview of successful and unsuccessful studies and implications for experimental design. J. Insect Physiol. 57, 231–245. https://doi.org/10.1016/j.jinsphys.2010.11.006.

Thakur, N., Mundey, J.K., Upadhyay, S.K., 2016. RNAi—implications in entomological research and pest control. In: Abdurakhmonov, I.Y. (Ed.), RNA Interference. InTech. https://doi.org/10.5772/61814.

Tiwari, S., Mann, R.S., Rogers, M.E., Stelinski, L.L., 2011. Insecticide resistance in field populations of Asian citrus psyllid in Florida. Pest Manag. Sci. 67, 1258–1268.

Tomich, J.M., Iwamoto, T., Hiromasa, Y., Gudlur, S. 2014. Branched amphipathic oligopeptides that self-assemble into vesicles. Patent No.:US 8,883,967 B2. United States Patent Publication. USPTO.

Tsai, J.H., Liu, Y.H., 2000. Biology of Diaphorina citri (Homoptera: Psyllidae) on four host plants. J. Econ. Entomol. 93, 1722–1725.

Turner, C.T., Davy, M.W., MacDiarmid, R.M., Plummer, K.M., Birch, N.P., Newcomb, R.D., 2006. RNA interference in the light brown apple moth, Epiphyas postvittana (Walker) induced by double-stranded RNA feeding. Insect Mol. Biol. 15, 383–391. https://doi.org/10.1111/j.1365-2583.2006.00656.x.

van den Berg, M.A., van Vuuren, S.P., Deacon, V.E., 1992. Studies on greening disease transmission by the citrus psylla, *Trioza erytreae* (Hemiptera: Triozidae). Isr. J. Entomol. 25–26, 51–56.

van der Zwet, T., Orolaza-Halbrendt, N., Zeller, W., 2012. Fire Blight History, Biology, and Management. American Phytopathological Soc. Press, St. Paul, MN.

Verma, A., 2018. Recent advances in antisense oligonucleotide therapy in genetic neuromuscular diseases. Ann. Indian Acad. Neurol. 21 (1), 3–8. https://doi.org/10.4103/aian.AIAN_298_17.

Wadia, J.S., Stan, R.V., Dowdy, S.F., 2004. Transducible TAT-HA fusogenic peptide enhances escape of TAT-fusion proteins after lipid raft macropinocytosis. Nat. Med. 10, 310–315. https://doi.org/10.1038/nm996.

Wang, W., Dai, H., Zhang, Y., Chandrasekar, R., Luo, L., Hiromasa, Y., Sheng, C., Peng, G., Chen, S., Tomich, J.M., Reese, J., Edwards, O., Kang, L., Reeck, G., Cui, F., 2015. Armet is an effector protein mediating aphid-plant interactions. FASEB J. 29, 2032–2045.

Wang, H., Russa, M.L., Qi, L.S., 2016. CRISPR/Cas9 in genome editing and beyond. Annu. Rev. Biochem. 85, 227–264. https://doi.org/10.1146/annurev-biochem-060815-014607.

Wang, N., Pierson, E.A., Setubal, J.C., Xu, J., Levy, J.G., Zhang, Y., Li, J., Rangel, L.T., Martins, J., 2017a. The Candidatus Liberibacter–host interface: insights into pathogenesis mechanisms and disease control. Annu. Rev. Phytopathol. 55, 20.1–20. https://doi.org/10.1146/annurev-phyto-080516-035513.

Wang, N., Stelinski, L.L., Pelz-Stelinski, K.S., Graham, J.H., Zhang, Y., 2017b. Tale of the huanglongbing disease pyramid in the context of the citrus microbiome. Phytopathology 107, 380–387. https://doi.org/10.1094/PHYTO-12-16-0426-RVW.

Wesolowski, D., et al., 2011. Basic peptide-morpholino oligomer conjugate that is very effective in killing bacteria by gene-specific and nonspecific modes. Proc. Natl. Acad. Sci. U. S. A. 108 (40), 16582–16587.

Wesolowski, D., Alonso, D., Altman, S., 2013. Combined effect of a peptide-morpholino oligonucleotide conjugate and a cell-penetrating peptide as an antibiotic. Proc. Natl. Acad. Sci. U. S. A. 110 (21), 8686–8689.

Whitten, M., Dyson, P., 2017. Gene silencing in non-model insects: overcoming hurdles using symbiotic bacteria for trauma-free sustainable delivery of RNA interference. Bioessays 39, 3. https://doi.org/10.1002/bies.201600247.

Whitten, M.M., Facey, P.D., Del Sol, R., Evans, M.C., Mitchell, J.J., Bodger, O.G., Dyson, P.J., 2016. Symbiont-mediated RNA interference in insects. Proc. R Soc. B. 283, 20160042. https://doi.org/10.1098/rspb.2016.0042.

Whyard, S., Singh, A.D., Wong, S., 2009. Ingested double-stranded RNAs can act as species-specific insecticides. Insect Biochem. Mol. Biol. 39, 824–832.

Whyard, S., Erdelyan, C.N., Partridge, A.L., Singh, A.D., Beebe, N.W., Capina, R., 2015. Silencing the buzz: a new approach to population suppression of mosquitoes by feeding larvae double-stranded RNAs. Parasit. Vectors 8, 96.

Wilson, R.C., Doudna, J.A., 2013. Molecular mechanisms of RNA interference. Annu. Rev. Biophys. 42, 217–239. https://doi.org/10.1146/annurev-biophys-083012-130404.

Wise, T.A. 2013. Can We Feed the World in 2050? A Scoping Paper to Assess the Evidence, Global Development and Environment Institute at Tufts University, Medford, MA, USA. http://www.ase.tufts.edu/gdae/Pubs/wp/13-04WiseFeedWorld2050.pdf.

Wise, T.A., Sundell, K., Brill, M., 2013. Rising to the challenge: changing course to feed the world in 2050. ActionAid USA Report, Oct. 2013. Global Development and Environment Institute at Tufts University, Medford, MA, USA. http://www.ase.tufts.edu/gdae/Pubs/rp/ActionAid_rising_to_challenge.pdf.

Wong, J.K.L., Mohseni, R., Hamidieh, A.A., MacLaren, R.E., Habib, N., Seifalian, A.M., 2017. Will nanotechnology bring new hope for gene delivery? Trends Biotechnol. 35, 434–451. https://doi.org/10.1016/j.tibtech.2016.12.009.

Wu, E.A., Terol, J., Ibanez, B., López-García, A., et al., 2018. Genomics of the origin and evolution of Citrus. Nature 544 (14), 311–316. https://doi.org/10.1038/nature25447.

Xu, C.F., Xia, Y.H., Li, K.B., Ke, C. 1988. Further study of the transmission of citrus huanglongbing by a psyllid, *Diaphorina citri* Kuwayama. Proc. 10th Conf. International Organization of Citrus Virologists, Nov. 17–21, 1986, Valencia, Spain (ed. by LW Timmer, SM Garnsey & L Navarro), pp. 243–248. IOCV, University of California, Riverside, CA, USA.

Yamamoto, P.T., Felippe, M.R., Sanches, A.L., Coelho, J.H.C., Garbim, L.F., Ximenes, N.L., 2009. Efficacy of insecticides for managing *Diaphorina citri* Kuwayama (Hemiptera: Psyllidae) in citrus. BioAssay 4, 1–9.

Ye, Y.H., Woolfit, M., Rancès, E., O'Neill, S.L., McGraw, E.A., 2013. Wolbachia-associated bacterial protection in the mosquito Aedes aegypti. PLoS Negl. Trop. Dis. 7, e2362.

Yoshida, T., Nakamura, H., Masutani, H., Yodoi, J., 2005. The involvement of thioredoxin and thioredoxin binding protein-2 on cellular proliferation and aging process. Ann. N. Y. Acad. Sci. 1055, 1–12. https://doi.org/10.1196/annals.1323.002. 16387713.

Youngblood, D.S., Hatlevig, S.A., Hassinger, J.N., Iversen, P.L., Moulton, H.M., 2007. Stability of cell-penetrating peptide-Morpholino oligomer conjugates in human serum and in cells. Bioconjug. Chem. 18, 50–60.

Younis, A., Siddique, M.I., Kim, C.K., Lim, K.B., 2014. RNA interference (RNAi) induced gene silencing: a promising approach of hi-tech plant breeding. Int. J. Biol. Sci. 10 (10), 1150–1158. https://doi.org/10.7150/ijbs.10452.

Yu, X., Siddarame, G., Killiny, N., 2017. Double stranded RNA delivery through soaking, mediates silencing of the muscle protein 20 and increases mortality to the Asian citrus psyllid, *Diaphorina citri*. Pest Manag. Sci. 73, 1846–1853. https://doi.org/10.1002/ps.4549.

Yuan, Y., Altman, S., 1994. Selection of guide sequences that direct efficient cleavage of mRNA by human ribonuclease P. Science 263, 1269–1273. https://doi.org/10.1126/science.8122108. PubMed: 8122108.

Zatsepin, T.S., Turner, J.J., Oretskaya, T.S., Gait, M.J., 2005. Conjugates of oligonucleotides and analogues with cell penetrating peptides as gene silencing agents. Curr. Pharm. Des. 11, 3639–3654. https://doi.org/10.2174/138161205774580769.

Zhang, L., Reed, R.D., 2017. A practical guide to CRISPR/Cas9 genome editing in Lepidoptera. bioRxiv. Preprint Apr.24. https://doi.org/10.1101/130344.

Zhang, X., Zhang, J., Zhu, K.Y., 2010. Chitosan/double-stranded RNA nanoparticle-mediated RNA interference to silence chitin synthase genes through larval feeding in the African malaria mosquito (*Anopheles gambiae*). Insect Mol. Biol. 19, 683–693.

Zotti, M.J., Smagghe, G., 2015. RNAi technology for insect management and protection of beneficial insects from diseases: lessons, challenges and risk assessments. Neotrop. Entomol. 44, 197–213. https://doi.org/10.1007/s13744-015-0291-8.

Zotti, M., Santos, E.A., Cagliari, D., Christiaens, O., Taning, C.N.T., Smagghe, G., 2018. RNA interference technology in crop protection against arthropod pests, pathogens and nematodes. Pest Manag. Sci. 74, 1239–1250. https://doi.org/10.1002/ps.4813.

FURTHER READING

Pan, W.-H., Clawson, G.A., 2006. Antisense applications for biological control. J. Cell. Biochem. 98, 14–35. https://doi.org/10.1002/jcb.20790.

Ramsey, J.S., Chavez, J.D., Johnson, R., Hosseinzadeh, S., et al., 2017. Protein interaction networks at the host–microbe interface in *Diaphorina citri*, the insect vector of the citrus greening pathogen. R. Soc. Open Sci. 4, 160545. https://doi.org/10.1098/rsos.160545.

Summerton, J., 1999. Morpholino antisense oligomers: the case for an RNase H-independent structural type. Biochim. Biophys. Acta 1489 (1), 141–158.

Waltz, E., 2016. Gene-edited CRISPR mushroom escapes US regulation. Nature 532 (7599), 293. https://doi.org/10.1038/nature.2016.19754.

Will, T., Vilcinskas, A., 2015. The structural sheath protein of aphids is required for phloem feeding. Insect Biochem. Mol. Biol. 57, 34–40. https://doi.org/10.1016/j.ibmb.2014.12.005.

Zhang, J., Khan, S.A., Hasse, C., Ruf, S., Heckel, D.G., Bock, R. 2015. Full crop protection from an insect pest by expression of long double-stranded RNAs in plastids. 347 (6225):991–994. https://doi.org/10.1126/science.1261680.

CHAPTER SEVEN

Screening, Efficacy and Mechanisms of Microbial Control Agents Against Sucking Pest Insects as Thrips

Shengyong Wu*^,2^, Liangde Tang†,2, Feng Fang‡, Donggang Li§, Xin Yuan¶, Zhongren Lei*, Yulin Gao*,1

*State Key Laboratory for Biology of Plant Disease and Insect Pests, Institute of Plant Protection, Chinese Academy of Agricultural Sciences, Beijing, PR China
†Environment and Plant Protection Institute, Chinese Academy of Tropical Agricultural Sciences, Haikou, PR China
‡Institute of Plant Protection, Shandong Academy of Agricultural Sciences, Jinan, PR China
§Dezhou Academy of Agricultural Sciences, Dezhou, PR China
¶Agricultural Experimental Station, Zhejiang University, Hangzhou, PR China
[1]Corresponding author: e-mail address: gaoyulin@caas.cn

Contents

Abstract

Polyphagous thrips are among the most important economically pests that cause serious damage in various ornamental and vegetable crops throughout the world. Because of their small size and cryptic habits, a number of thrips species are easy to invade into

[2] These authors contributed equally to this work.

Advances in Insect Physiology, Volume 55
ISSN 0065-2806
https://doi.org/10.1016/bs.aiip.2018.07.005

new areas. For a long period, thrips control has mainly relied on frequent use of insecticides, which lead to a series of ecological problems. Although a number of alternative management tactics have been developed in many cropping systems, many invasive thrips continue to spread internationally and display vast damage potential. This chapter reviews microbial control of thrips with focus on the entomopathogenic fungi *Beauveria bassiana, Metarhizium anisopliae, Isaria fumosorosea* and *Lecanicillium lecanii.* The strains of fungal pathogens that were screened and proven to be effective for control of thrips should be developed worldwide and available for growers. We also discuss three areas of future research that will be necessary in improving management programs of thrips.

1. INTRODUCTION

Thrips is the general term of Thysanoptera. They are often considered as one of the most economically important insect pests in agroecosystems all over the world (Lewis, 1973; Riley et al., 2018). More than 7700 species were recorded, despite only about 1% of this group could damage the various crops (Morse and Hoddle, 2006), but the importance is hard to overestimate, especially for vector thrips of plant viruses (Riley et al., 2011; Rotenberg et al., 2015). In these thrips, some species are global pests, e.g., *Frankliniella occidentalis* (Pergande) (Kirk and Terry, 2003; Reitz, 2009), *Thrips tabaci* (Lindeman) (Diaz-Montano et al., 2011) and *Thrips palmi* (Karny) (Murai and Loomans, 2001; Rosenheim et al., 1990), some species are regional distribution, e.g., *Thrips hawaiiensis* (Morgan) (Murai, 2001). However, some species have broad range of hosts, for instance, the species in genus of *Frankliniella* (Northfield et al., 2008; Reitz et al., 2003) and some species are oligophagous, such as *Megalurothrips sjostedti* (Trybom) (Nabirye et al., 2003; Ngakou et al., 2008; Tamò et al., 1993) and *Megalurothrips usitatus* (Bagnall) (Chang, 1987, 1988; Tang et al., 2015) major feeding on leguminous plants. In a summary, these thrips are thus as key pests because they have the same characteristics of wide distribution or wide host range, serious harmful and large yield losses.

Thrips usually have high fecundity and short developmental time, which results in serious overlapping generation, high infestation frequency and quite resistance to most insecticides (Lewis, 1997; Morse and Hoddle, 2006). It can cause tremendous damage to crops, both by feeding and oviposition. The traits of their piercing–sucking mouthparts, which enable them to feed on different types of surface tissue cells. Feeding on actively growing tissue leads to distortion, poorly expanded and eventually death, which cause yield losses, while feeding on expanded tissue results in the characteristic silvery-white spots, which cause aesthetic damage and reduce market value (Annamalai et al., 2016; Shipp et al., 1998). Largely economic losses exceed direct damage if thrips serve as a vector of plant viruses (Riley et al., 2011; Rotenberg et al., 2015).

A serial of management tactics for thrips have been explored, the most common measure is use chemicals. Owing to most thrips nymphs and adults live inside narrow spaces, such as flowers, buds, young unfold leaves or leaves blade abaxially, where spray coverage is difficult to reach, consequently lead to a poor control efficacy (Loughner et al., 2005). In addition, insecticide resistance is a severe problem for thrips control. For example, *F. occidentalis* has developed resistance to insecticides in several parts of the world (Bielza et al., 2010; Immaraju et al., 1992; Wang et al., 2011a, b; Zhao et al., 1995). *T. tabaci* (Herron et al., 2008; Macintyre Allen et al., 2005; Martin et al., 2003; Morishita, 2008; Shelton et al., 2006) and *T. palmi* (Bao et al., 2014; Bao and Sonoda, 2012) are faced with the same resistance problems. Crop resistance may offer a long-term solution to thrips control and reduce the use of insecticides (Diaz-Montano et al., 2011; Kennedy, 2008). Although it is recommended as a viable strategy by some published studies (Abudulai et al., 2006; Alabi et al., 2003; Diaz-Montano et al., 2010; Romero-González et al., 2010), it is difficult to come true in recent years because imitation of commercial resistant cultivars. Physical control, e.g., sticky traps with colours, is used to reduce the population of thrips in greenhouses and open fields. Despite most of thrips species were sensitive to one specific colour (Tang et al., 2016), the practical effect is inferior and thus the stick traps have usually been used to monitor thrips (Jacobson, 1997; Pizzol et al., 2010) or as an accompany measure. One another major method is using biological agents, including natural enemies and entomopathogenic fungus. Biological control efforts have focused on using predators, such as predatory bugs in the genus *Orius* (Castañé et al., 1999; Cox et al., 2006; Kawai, 1995; Riudavets, 1995; van Lenteren et al., 1997) and predatory mites in the genera *Amblyseius* and *Hypoaspis* (Cox et al., 2006; Messelink et al., 2005; Murphy et al., 1998), and have been largely unsuccessful. This is usually due to the application of insecticides that are harmful to predators. However, entomopathogenic fungi have been used as potential biological control agent for some thrips species (Murphy et al., 1998; Thungrabeab et al., 2006). In this review, we target to the research and application status, pathogenic mechanism and future directions for research of entomopathogenic fungi on thrips control.

2. SPECIES OF MICROBIAL CONTROL AGENTS AGAINST INSECT THRIPS

Microbial agents used for thrips control mainly belong to *Beauveria bassiana*, *Metarhizium anisopliae*, *Isaria fumosorosea*, *Lecanicillium lecanii* (formerly Verticillium), etc. (Table 1). Although these microbial control agents have a broad host range, species and even fungal isolates within a species usually behave very differently. Therefore, effective strains of fungal pathogens against the target host should be screened and evaluated based on laboratory bioassay and field evaluation.

2.1 *B. bassiana*

B. bassiana is widely used for thrips control. Laboratory experiments showed that *B. bassiana* were high pathogenic to various species of thrips, including *T. tabaci* (Abe and Ikegami, 2005; Annamalai et al., 2013; Ezzati-Tabrizi et al., 2009; Wu et al., 2013), *F. occidentalis* (Gao et al., 2012a, b; Sengonca et al., 2006; Wu et al., 2015) and *Gynaikothrips uzeli* (Zimmerman) (Sánchez-Peña et al., 2011). Because of differences in virulence between strains, Wu et al. (2014) tested 28 fungal strains by bioassays and screened *B. bassiana* LNSZ-26 that is the most virulent strain, cause 96% mortality of *F. occidentalis*. This strain also significantly reduced *F. occidentalis* populations on greenhouse cucumber, with a decreased rate of 80% and 78% in the adult and larval population, respectively (Wu et al., 2016). Field/greenhouse trials demonstrated that *B. bassiana* displayed high efficacy to control *T. tabaci* on onion with commercial formulation (Myco-Jaal®) (Annamalai et al., 2016) and on tomato plants (Almazraawi et al., 2009), and significantly suppressed *T. palmi* population on potato crop (Silva et al., 2011), green bean plants (Castineiras et al., 1996) and melon plants (Saito, 1991). Similar results were found in *F. occidentalis* on cucumber (Wu et al., 2016, 2017), impatiens crops (Ugine et al., 2007) and French marigolds (Skinner et al., 2012). *B. bassiana* (BotaniGard®) applied at weekly intervals can also significantly reduced numbers of *Scirtothrips dorsalis* adults and larvae on pepper (Seal et al., 2007). *Scirtothrips citri* (Moulton) and *Thrips parvispinus* (Karny) populations were also significantly reduced on blueberries (Zahn et al., 2013) and on chrysanthemum (Silvia Yusuf et al., 2010) after application of *B. bassiana*, respectively. The findings of *B. bassiana* oil formulation against thrips of okra were on par with chemical treatment (Oxydemeton methyl 25EC), which showed excellent field

Table 1 Biological Control With Differential Entomopathogenic Fungi Against Various Species Thrips

Fungal Species	Target Thrips Species	References
Beauveria bassiana	*Thrips tabaci* Lindeman	Saito (1991), Thungrabeab et al. (2006), Almazraawi et al. (2009), Ezzati-Tabrizi et al. (2009), Wu et al. (2013) and Annamalai et al. (2013, 2016)
	Frankliniella occidentalis Pergande	Gill et al. (1998), Murphy et al. (1998), Sengonca et al. (2006), Ugine et al. (2007), Gao et al. (2012a, b), Skinner et al. (2012) and Wu et al. (2015, 2016, 2017)
	Frankliniella invasor Sakimura	Toledo-Hernández et al. (2017)
	Megalurothrips sjostedti Trybom	Ekesi et al. (1998a, b)
	Thrips palmi Karny	Saito (1991), Castineiras et al. (1996) and Silva et al. (2011)
	Scirtothrips dorsalis Hood	Seal et al. (2007) and Arthurs et al. (2013)
	Scirtothrips citri Moulton	Zahn et al. (2013)
	Gynaikothrips uzeli Zimmerman	Sánchez-Peña et al. (2011)
Metarhizium anisopliae/brunneum	*Frankliniella occidentalis*	Vestergaard et al. (1995), Brownbridge (1995), Lopes et al. (2000), Maniania et al. (2002), Azaizeh et al. (2002), Sengonca et al. (2006), Ansari et al. (2007, 2008), Skinner et al. (2012), Niassy et al. (2012), Koupi et al. (2016) and Dura et al. (2012)
	Frankliniella invasor Sakimura	Toledo-Hernández et al. (2017)
	Megalurothrips sjostedti Trybom	Ekesi et al. (1998a, b, 1999) and Ekesi and Maniania (2000)
	Scirtothrips dorsalis Hood	Seal et al. (2007) and Arthurs et al. (2013)

Continued

Table 1 Biological Control With Differential Entomopathogenic Fungi Against Various Species Thrips—cont'd

Fungal Species	Target Thrips Species	References
	Thrips tabaci Lindeman	Maniania et al. (2003) and Thungrabeab et al. (2006)
	Thrips palmi Karny	Iwase and Shimizu (2004) and Silva et al. (2011)
Lecanicillium (*Verticillium*) *lecanii*/*muscarium*	*Thrips tabaci* Lindeman	Annamalai et al. (2016) and Gillespie (1986)
	Thrips palmi Karny	Saito (1992), North et al. (2006) and Silva et al. (2011)
	Frankliniella occidentalis	Van der Schaaf et al. (1990), Heyler (1993) and Vestergaard et al. (1995)
	Taeniothrips inconsequens Uzel	Skinner et al. (1991)
Isaria fumosorosea	*Scirtothrips dorsalis* Hood	Arthurs et al. (2013)
Entomophthora	*Thrips tabaci* Lindeman	Carl (1975), Macleod et al. (1976) and Samson et al. (1979)
	Thrips palmi Karny	Saito et al. (1989)
	Frankliniella occidentalis	Vacante et al. (1994)
Peacilomyces	*Thrips tabaci* Lindeman	Thungrabeab et al. (2006)

efficacy (Naik and Shekharappa, 2009). However, early research indicated that some strains of *B. bassiana* were not effective to onion thrips on onion (Gerace et al., 1996) and thrips on pepper plants (Frantz and Mellinger, 1998).

2.2 *M. anisopliae*

Biological control of thrips with *M. anisopliae* has shown excellent potential for management of thrips in various crop ecosystems. Four isolates of *M. anisopliae* from Thailand were found to be high pathogenicity level to

F. occidentalis (Sengonca et al., 2006). The isolate of *M. anisopliae* (DEMI002) from Iranian showed a high virulence in *F. occidentalis* (Koupi et al., 2016), and the isolate 275 resulted in at least 94% mortality to *F. occidentalis* (Vestergaard et al., 1995), the commercial formulation Meta-TNK® also showed high pathogenicity (91%) to *Frankliniella invasor*, which belong to the genus of *Frankliniella* (Toledo-Hernández et al., 2017). Iwase and Shimizu (2004) reported that the virulence of *M. anisopliae* to *T. palmi* is correlated with the number of adhered conidia on thrips body, and showed low LC_{50} values in isolate of OE-2. The fungus can also significantly reduced by both the adult and larval populations of *F. occidentalis* on chrysanthemum in greenhouse. Soil applications of *M. anisopliae* could reduce *F. occidentalis* population significantly and prevent major damage to their host plants (Skinner et al., 2012). The results obtained from Maniania et al. (2003) indicated that the potential of using *M. anisopliae* for the control of *T. tabaci* while protecting biodiversity in the onion agroecosystem in Kenya. Dura et al. (2012) found that *M. anisopliae* F52 was as effective as the chemical insecticides against *F. occidentalis* on pepper plant, while Ansari et al. (2007) considered *M. anisopliae* V275 was more efficacious than chemical insecticides (imidacloprid and fipronil) for control pupae of *F. occidentalis* after treated with growing media, which offers much promise for control of thrips as part of an integrated pest management (IPM) program. Similar results were observed on *M. anisopliae* against *F. occidentalis* on hidroponic lettuce production (Lopes et al., 2000). Field experiments showed that *M. anisopliae* had a good efficacy for control of *M. sjostedti* on cowpea, and a better efficacy was found if intercrop with maize (Ekesi et al., 1999). The results of *M. anisopliae* effectively reduced *S. dorsalis* adults on cotton, which provides excellent control of this pest (Seal et al., 2007). Combined with semiochemical, *M. anisopliae* still keep a good performance for thrips control (Niassy et al., 2012).

2.3 *L. lecanii*

L. lecanii is another effective entomopathogenic fungus against thrips pests. High efficacy of *L. lecanii* was showed in the field/greenhouses experiment against *T. palmi* (North et al., 2006; Saito, 1992; Silva et al., 2011). The fungus also caused high mortality to *T. tabaci* (Annamalai et al., 2013) and a range of extent mortality (20%–70%) to *F. occidentalis* (Vestergaard et al., 1995). In addition, good efficacy was exhibited after application of *L. lecanii* (Phule Bugicide®) to control *T. tabaci* on onion plants under greenhouse as well as field conditions (Annamalai et al., 2016). Field evaluation also demonstrated *V.* (=*L.*) *lecanii* was an available tool for thrips management on okra (Naik and Shekharappa, 2009) and can be considered as a promising tool in biological control programs of the pear thrips *Taeniothrips inconsequens* Uzel in forest habitats of Vermont (Skinner et al., 1991).

2.4 Other Entomopathogenic Fungi

Besides of mainly entomopathogenic fungi listed earlier, some other microbial control agents, such as *I. fumosorosea* (Arthurs et al., 2013) and some strains in the genus of *Entomophthora* (Carl, 1975; Macleod et al., 1976; Saito et al., 1989; Samson et al., 1979; Vacante et al., 1994) and *Peacilomyces* (Thungrabeab et al., 2006), also demonstrated extent potential to control various species of thrips in different crop ecosystems.

3. MECHANISMS OF MICROBIAL CONTROL AGENTS AGAINST INSECT PESTS

Asexually produced fungal conidia or spores are generally responsible for infection to host insects (Shah and Pell, 2003). Entomopathogenic fungus, such as *B. bassiana* and *M. anisopliae*, successfully infect host undergo the following stages: (1) conidial adhesion on the cuticle of suitable host, (2) conidial germination and germ tube elongation, (3) penetration of insect cuticle, (4) spread of hyphal bodies within host and (5) sporulation on the internal surfaces of the desiccated host. During this process, the fungus produces many specialized infection structures including appressoria and/or penetration pegs, which enable the growing hyphae to penetrate the host integument. In the later stages of this process, the fungus overcomes a serial of internal immune responses of the host (Ortiz-Urquiza and Keyhani, 2013). Nutrients acquired from host by hyphal bodies, mechanical disruption of tissues by mycelial growth and secondary metabolites production in the insect hemolymph finally lead to the host death within 1 week after infection. On death of insect hosts, fungal mycelia grow from the dead body and sporulation usually occurs on the ventral surface of the cadaver. Conidia are dispersed throughout the environment and further transmit to the new hosts.

When sprayed with *B. bassiana* to adult *F. occidentalis*, a great deal of conidia could adhere to its cuticle, including antenna, compound eyes, folded regions of legs, setal alveolus, wings, ovipositor and the joints of arthromere (Wang et al., 2011a, b). It was reported that conidia of *B. bassiana* strain LNSZ-26 that is highly virulent to *F. occidentalis* attach its cuticle within 2 h after inoculation and germinated oriented towards the cuticle within 24 h. Germ tubes of conidia successfully penetrated the thrips cuticle within 48 h, with fungal mycelia prolifically growing from dead adults within 72 h (Wu et al., 2014). Observations made on aberrant behaviour of the legume flower thrips, *M. sjostedti* (Trybom) after inoculated with *M. anisopliae* showed leg twitching and abdominal arching of adults within 3–4 days (Ekesi and Maniania, 2000). However, immature stages of thrips are generally less susceptible to fungal infection than the adult stage because moulting avoids contact with fungal inoculum (Mouden et al., 2017; Ugine et al., 2005).

In addition to the direct pathogenicity, sublethal effects of fungus to thrips including changes in development time, feeding, fecundity, fertility and longevity, which play an important role in determining the population dynamics, may be one of the most important aspects of exposure to microbial control agents. Ekesi and Maniania (2000) reported that *M. anisopliae* can result to a significant reduction in fecundity, egg fertility and longevity of *M. sjostedti* adult. Zhang et al. (2015) reported that a strain of *B. bassiana* GZGY-1-3 significantly reduced the reproductive success of *F. occidentalis* and the subsequent fitness of their progeny. Similarly, Ugine et al. (2006) reported that *B. bassiana* can significantly decreased thrips longevity, the ovipositional period and lifetime fecundity. It is possible that *B. bassiana* competes with the host for nutrients and subsequently physiologically interfere with the egg fertility. Therefore, this sublethal effects could partly offset the relatively slow speed of death by fungal infection compared to conventional insecticides and may play an important role in thrips control.

4. FUTURE PROSPECTS

4.1 Formulations Development Targeting the Soil-Dwelling Stage of Thrips

An ideal biological control strategy would target both the foliar-feeding and soil-dwelling development stages of the pest (Ansari et al., 2008). To date, microbial control of thrips has largely focused on adults and larvae in the crop canopy. However, this approach was not always successful, because the majority of thrips leave the host plants at late second instars and pupate in the soil before adult emergence (Berndt et al., 2004; Hulshof et al., 2003). According to Berndt (2002), *F. occidentalis* spend more than one-third of its developmental cycle for pupation in the soil. Effort should be made to interrupt the life cycle by controlling the pupae, and fungal-based granular formulations should be developed in the future. Because relatively high humidity of soil offer a more conducive environment, which improves survival of fungal conidia and enhances the germination, fungal proliferation will increase viable inoculum and efficacy in thrips control (Lee et al., 2017; Skinner et al., 2012). Therefore, targeting soil-dwelling stages of thrips will be an effective strategy to manage thrips in field.

4.2 Evaluation of Compatibility With Other Biological Control Agents

Until recently, a number of fungal pathogens have been developed worldwide for the control of various species of thrips. Because of a broad host range, arthropod pathogens often exert a negative effect on the predators and may consequently disrupt pest control (Ong and Vandermeer, 2015; Thungrabeab and Tongma, 2007). Therefore, the success of pathogens as biological control agents depends not only on their high efficacy against pests but also on safety to beneficial arthropods, such as natural enemies of insect pests (Shipp et al., 2012; Stuart et al., 2011). Roy and Pell (2000) reported that some species of fungal pathogens may have direct negative effects on predators if infected hosts were less appropriate as prey. Seiedy et al. (2015) reported that three strains of *B. bassiana* (DEBI008, F and J.B) displayed high pathogenicity to *Amblyseius swirskii* Athias-Henriot, a high effective predator of thrips. Therefore, evaluating the compatibility of the potential pathogens and beneficial arthropods may be critical to the success of thrips control. For example, Gao et al. (2012b) found a strain of *B. bassiana* Bb-RSB displayed highly virulence against *F. occidentalis* without negative effects on its natural enemy *Orius sauteri* (Poppius). In addition, commonly laboratory based with bioassay experiments being conducted in Petri dishes to evaluate and predict the effects on natural enemies are not readily applicable to agroecosystems, complementary studies, should be carried out in greenhouse or field conditions.

4.3 Improving the Resistance of Fungal Strains

Most formulations of fungal pathogens are available for foliar application, which is a primary control method used by growers. However, the efficacy has always been inconsistent, because fungal conidia sprayed on leaves will dry out in a short time and loss vitality or are inactivated by ultraviolet light from sunlight and high temperature. It was reported that conidial germination was severely impaired at 38 and 40°C for all *Metarhizium* species (Rangel et al., 2010). Giving the effects of external conditions on fungal conidia, insect pathogenic fungi used has been mainly limited in a few greenhouse crops (Ugine et al., 2007; Wu et al., 2013). Therefore, high resistance of fungal strains targeting thrips need to be screened for wider commercial application. Additional measures will also contribute to improve the fungal resistance. These include: (a) to optimize the concentrations of conidial heat-protective saccharides for improving the efficacy of conidial germination rate of *B. bassiana* (Li and Lee, 2014); (b) to amend plant oil as medium for increasing fungal conidial thermotolerance (Kim et al., 2010) and (c) to release more glucose as substrates for enhancing conidial thermotolerance and field persistence (Ying and Feng, 2006). In addition, new biotechnology, i.e., genetic manipulations should be explored in fungal pathogens to enhance their environmental tolerance. Leng et al. (2011) reported that genetically engineering the trehalose pathway would increase the thermotolerance of fungal strains without compromising their virulence, which offer a new strategy for enhancing conidiospore thermotolerance of fungal pathogens.

5. CONCLUSIONS

Successful microbial control of crop pests will ultimately depend on how carefully the fungal strains are screened (virulence, stress tolerance and persistence), on the scale of the production, formulation, compatibility with other biological control agents and on the efficacy of the delivery system, and the timing of the application. Although many isolates of fungal pathogens have been found to be highly virulent against thrips, very few strains, such as *B. bassiana* GHA (Botanigard®, Laverlam International and Naturalis-L®, Troy BioSciences) were registered as a commercial products (Cloyd and Sadof, 2003; Jacobson et al., 2001). Growers are more hopeful to see more fungal strains registered and cost-effective products available for use against thrips, including greenhouse and outdoor plantings. In addition, because of many factors affecting the efficacy of fungal insect pathogens, efforts should be taken to ensure that management programs of thrips are cost effective and easy for growers to implement.

FUNDING

This work was supported by National Key Research and Development Project (Grant No. 2016YFC1201200) and Beijing Natural Science Foundation (Grant No. 6172031).

REFERENCES

Abe, M., Ikegami, T., 2005. Susceptibility of five species of thrips to different strains of the entomopathogenic fungus, *Beauveria bassiana*. Appl. Entomol. Zool. 40 (4), 667–674.

Abudulai, M., Salifu, A.B., Haruna, M., 2006. Screening of cowpeas for resistance to the flower bud thrips, *Megalurothrips sjostedti* Trybom (Thysanoptera: Thripidae). J. Appl. Sci. 6 (7), 1621–1624.

Alabi, O.Y., Odebiyi, J.A., Jackai, L.E.N., 2003. Field evaluation of cowpea cultivars (*Vigna unguiculata* [L.] Walp.) for resistance to flower bud thrips (*Megalurothrips sjostedti* Trybom) (Thysanoptera: Thripidae). Int. J. Pest Manage. 49 (4), 287–291.

Almazraawi, M.S., Al-Abbadi, A., Shatnawi, M.A., Ateyyat, M., 2009. Effect of application method on the interaction between *Beauveria bassiana* and neem tree extract when combined for *Thrips tabaci* (Thysanoptera: Thripidae) control. J. Food Agric. Environ. 7 (2), 869–873.

Annamalai, M., Kaushik, H.D., Selvaraj, K., 2013. Pathogenicity of *Beauveria bassiana* (Balsamo) *Vuillemin* and *Lecanicillium lecanii* Zimmerman against onion thrips, *Thrips tabaci* Lindeman. Biopestic. Int. 9 (2), 148–156.

Annamalai, M., Kaushik, H.D., Selvaraj, K., 2016. Bioefficacy of *Beauveria bassiana* (Balsamo) Vuillemin and *Lecanicillium lecanii* Zimmerman against *Thrips tabaci* Lindeman. Proc. Natl. Acad. Sci. India Sect. B Biol. Sci 86 (2), 505–511. https://doi.org/10.1007/s40011-014-0475-8.

Ansari, M.A., Shah, F.A., Whittaker, M., Prasad, M., Butt, T.M., 2007. Control of western flower thrips (*Frankliniella occidentalis*) pupae with *Metarhizium anisopliae* in peat and peat alternative growing media. Biol. Control. 40, 293–297.

Ansari, M.A., Brownbridge, M., Shah, F.A., Butt, T.M., 2008. Efficacy of entomopathogenic fungi against soil-dwelling life stages of western flower thrips, *Frankliniella occidentalis*, in plant-growing media. Entomol. Exp. Appl. 127 (2), 80–87.

Arthurs, S.P., Aristizábal, L.F., Avery, P.B., 2013. Evaluation of entomopathogenic fungi against chilli thrips, *Scirtothrips dorsalis*. J. Insect Sci. 13 (31), 1–16.

Azaizeh, H., Gindin, G., Said, O., Barash, I., 2002. Biological control of the western flower thrips *Frankliniella occidentalis* in cucumber using the entomopathogenic fungus *Metarhizium anisopliae*. Phytoparasitica 30 (1), 18–24.

Bao, W.X., Sonoda, S., 2012. Resistance to cypermethrin in melon thrips, Thrips palmi (Thysanoptera: Thripidae), is conferred by reduced sensitivity of the sodium channel and CYP450-mediated detoxification. Appl. Entomol. Zool. 47 (4), 443–448.

Bao, W.X., Narai, Y., Nakano, A., Kaneda, T., Murai, T., Sonoda, S., 2014. Spinosad resistance of melon thrips, *Thrips palmi*, is conferred by g275e mutation in α6 subunit of nicotinic acetylcholine receptor and cytochrome p450 detoxification. Pestic. Biochem. Physiol. 112 (1), 51–55.

Berndt, O., 2002. Entomopathogenic nematodes and soil dwelling predatory mites: suitable antagonists for enhanced biological of *Frankliniella occidentalis* (Pergande) (Thysanoptera: Thripididae)? PhD thesisUniversity of Hannover, Hanover.

Berndt, O., Meyhöfer, R., Poehling, H.M., 2004. The edaphic phase in the ontogenesis of *Frankliniella occidentalis* and comparison of Hypoaspis miles and Hypoaspis aculeifer as predators of soildwelling thrips stage. Biol. Control 30, 17–24.

Bielza, P., Quinto, V., Contreras, J., Torné, M., Martín, A., Espinosa, P.J., 2010. Resistance to spinosad in the western flower thrips, *Frankliniella occidentalis* (pergande), in greenhouses of South-Eastern Spain. Pest Manag. Sci. 63 (7), 682–687.
Brownbridge, M., 1995. Prospect for mycopathogens in thrips management. In: Parker, M., Skinner, M., Lewis, T. (Eds.), Thrips Biology and Management. Plenum Press, New York, pp. 281–295.
Carl, K.P., 1975. An Entomophthora sp. (Entomophthorales: Entomophthoraceae) pathogenic to Thyips spp. (Thysanoptera: Thripidae) and its potential as a biological control agent in glasshouses. Entomophaga 20, 381–388.
Castañé, C., Riudavets, J., Yano, E., 1999. Biological control of thrips. In: Albajes, R., Gullino, M.L., van Lenteren, J.C., Elad, Y. (Eds.), Integrated Pest and Disease Management in Greenhouse Crops. Kluwer Academic, Dordrecht, pp. 244–253.
Castineiras, A., Peña, J.E., Duncan, R., Osborne, L., 1996. Potential of *Beauveria bassiana* and *Paecilomyces fumosoroseus* (deuteromycotina: hyphomycetes) as biological control agents of *Thrips palmi* (Thysanoptera: Thripidae). Fla Entomol. 79 (3), 458–461.
Chang, N.T., 1987. Seasonal abundance and developmental biology of thrips *Megalurothrips usitatus* on soybean at southern area of Taiwan. Plant Prot. Bull. 29 (2), 165–173.
Chang, N.T., 1988. The preference of thrips, *Megalurothrips usitatus* (Bagnall), for three leguminous plants. Plant Prot. Bull. 30 (1), 68–77.
Cloyd, R.A., Sadof, C.S., 2003. Western flower thrips. Ornementals & Turf E-110-W. IN, USA, Department of Entomology, Purdue University, p. 4.
Cox, P.D., Matthews, L., Jacobson, R.J., Cannon, R., MacLeod, A., Walters, K.F.A., 2006. Potential for the use of biological agents for the control of *Thrips palmi* (Thysanoptera: Thripidae) outbreaks. Biocontrol Sci. Dent. Technol. 16 (9), 871–891.
Diaz-Montano, J., Fuchs, M., Nault, B.A., Shelton, A.M., 2010. Evaluation of onion cultivars for resistance to onion thrips (Thysanoptera: Thripidae) and iris yellow spot virus. J. Econ. Entomol. 103, 925–937.
Diaz-Montano, J., Fuchs, M., Nault, B.A., Fail, J., Shelton, A.M., 2011. Onion Thrips (Thysanoptera: Thripidae): a global pest of increasing concern in onion. J. Econ. Entomol. 104 (1), 1–13.
Dura, O., Dura, S., Bilen, E., 2012. Determination of the effective dose of *Metarhizium anisopliae* str. F 52 for biological control of western flower thrips on pepper plant. Bahçe 41 (1), 1–7.
Ekesi, S., Maniania, N.K., 2000. Susceptibility of *Megalurothrips sjostedti* developmental stages to *Metarhizium anisopliae* and the effect of infection on feeding, adult fecundity, egg fertility and longevity. Entomol. Exp. Appl. 94, 229–236.
Ekesi, S., Maniania, N.K., Ampong-Nyarko, K., Onu, I., 1998a. Potential of the entomopathogenic fungus *Metarhizium anisopliae* (Metsch.) Sorok. For the control of the legume flower thrips, *Megalurothrips sjostedti* (Trybom) on cowpea in Kenya. Crop Prot. 17, 661–668.
Ekesi, S., Maniania, N.K., Onu, I., Löhr, B., 1998b. Pathogenicity of entomopathogenic fungi (Hyphomycetes) to the legume flower thrips, *Megalurothrips sjostedti* (Trybom) (Thys., Thripidae). J. Appl. Entomol. 122, 629–634.
Ekesi, S., Maniania, N.K., Ampong-nyarko, K., Onu, I., 1999. Effect of intercropping cowpea with maize on the performance of *Metarhizium anisopliae* against *Megalurothrips sjostedti* (Thysanoptera: Thripidae) and predators. Environ. Entomol. 28 (6), 1154–1161.
Ezzati-Tabrizi, R., Talaei-Hassanloui, R., Pourian, H.R., 2009. Effect of formulating of *Beauveria bassiana* conidia on their viability and pathogenicity to the onion thrips, *Thrips tabaci* lind. (thysanoptera: thripidae). J. Plant Prot. Res. 49 (1), 97–104.
Frantz, G., Mellinger, H.C., 1998. Potential use of *Beauveria bassiana* for biological control of thrips in peppers. Proc. Fla. State Hort. Soc. 111, 79–81.

Gao, Y.L., Reitz, S.R., Wang, J., Xu, X.N., Lei, Z.R., 2012a. Potential of a strain of the entomopathogenic fungus *Beauveria bassiana* (Hypocreales: Cordycipitaceae) as a biological control agent against western flower thrips, *Frankliniella occidentalis* (Thysanoptera: Thripidae). Biocontrol Sci. Technol. 22 (4), 491–495.

Gao, Y.L., Reitz, S.R., Wang, J., Tamez-Guerra, P., Wang, E.D., Xu, X.N., Lei, Z.R., 2012b. Potential use of the fungus *Beauveria bassiana* against the western flower thrips *Frankliniella occidentalis* without reducing the effectiveness of its natural predator *Orius sauteri* (Hemiptera: Anthocoridae). Biocontrol Sci. Dent. Technol. 22 (7), 803–812.

Gerace, D., Sclar, D.C., Cranshaw, W., 1996. Control of onion thrips using *Beauveria bassiana* formulations. Arthropod Manag. Tests 22, 144.

Gill, S.A., Reeser, R., Raupp, M.J., 1998. Battling thrips: five pesticides put to the test. Grower Talks 10, 46–48.

Gillespie, A.T., 1986. In: The potential of entomogenous fungi as control agents for onion thrips, *Thrips tabaci*. Proceeding of Brighton Crop Protection Conference: Biotechnology and Crop Improvement and Protection, Brighton, UK, pp. 237–242.

Herron, G.A., James, T.M., Mo, J.H., 2008. Australian populations of onion thrips, *Thrips tabaci* Lindeman (Thysanoptera: Thripidae), are resistant to some insecticides used for their control. Aust. J. Entomol. 47, 361–364.

Heyler, N.L., 1993. Verticillium lecanii for the control of aphids and thrips on cucumber. IOBC/WPRS Bull. 16 (2), 63–66.

Hulshof, J., Ketoja, E., Vänninen, I., 2003. Life history characteristics of *Frankliniella occidentalis*, on cucumber leaves with and without supplemental food. Entomol. Exp. Appl. 108 (1), 19–32.

Immaraju, J.A., Paine, T.D., Bethke, J.A., Robb, K.L., Newman, J.P., 1992. Wester flower thrips (Thysanoptera: Thripidae) resistance to insecticides in coastal California greehouses. J. Econ. Entomol. 85 (1), 9–14.

Iwase, R., Shimizu, S., 2004. Adhesion and virulence of conidia of *Metarhizium anisopliae* to *Thrips palmi* (thysanoptera: thripidae). Jpn J. Appl. Entomol. Zool. 48 (4), 275–280.

Jacobson, R.J., 1997. Integrated pest management (IPM) in glasshouses. In: Lewis, T. (Ed.), Thrips as Crop Pests. CAB International, New York, pp. 639–666.

Jacobson, R.J., Chandler, D., Fenlon, J., Russell, K.M., 2001. Compatibility of *Beauveria bassiana* (Balsamo) Vuillemin with *Amblyseius cucumeris* Oudemans (Acarina: Phytoseiidae) to control *Frankliniella occidentalis* Pergande (Thysanoptera: Thripidae) on cucumber plants. Biocontrol Sci. Dent. Technol. 11 (3), 391–400.

Kawai, A., 1995. Control of *Thrips palmi* Karny (Thysanoptera: Thripidae) by *Orius* spp. (Heteroptera: Anthocoridae) on greenhouse eggplant. Appl. Entomol. Zool. 30 (1), 1–6.

Kennedy, G.G., 2008. Integration of insect-resistant genetically modified crops within IPM programs. In: Romeis, J., Shelton, A.M., Kennedy, G.G. (Eds.), Integration of Insect-Resistant Genetically Modified Crops Within IPM Programs. Springer, Dordrecht, The Netherlands, pp. 1–26.

Kim, J.S., Skinner, M., Parker, B., 2010. Plant oils for improving thermotolerance of *Beauveria bassiana*. J. Microbiol. Biotechnol. 20 (9), 1348–1350.

Kirk, W.D., Terry, I., 2003. The spread of the western flower thrips *Frankliniella occidentalis* (Pergande). Agric. For. Entomol. 5, 301–310.

Koupi, N., Ghazavi, M., Kamali, K., Baniameri, V., 2016. Virulence of Iranian isolates of *Metarhizium anisopliae* on western flower thrips, *Frankliniella occidentalis* (Pergande) (Thysanoptera: Thripidae). J. Crop Prot. 5 (1), 131–138.

Lee, S.J., Kim, S., Kim, J.C., Mi, R.L., Hossain, M.S., Shin, T.S., Kim, T.H., Kim, J.S., 2017. Entomopathogenic *Beauveria bassiana*, granules to control soil-dwelling stage of western flower thrips, *Frankliniella occidentalis*, (thysanoptera: thripidae). Biocontrol 62 (5), 1–10.

Leng, Y.J., Peng, G.X., Cao, Y.Q., Xia, Y.X., 2011. Genetically altering the expression of neutral trehalase gene affects conidiospore thermotolerance of the entomopathogenic fungus *Metarhizium acridum*. BMC Microbiol. 11, 32.

Lewis, T., 1973. Thrips: Their Biology, Ecology and Economic Importance. Academic Press, London.

Lewis, T., 1997. Pest thrips in perspective. In: Lewis, T. (Ed.), Thrips as Crop Pests. CAB International, Wallingford, UK, pp. 1–14.

Li, H.X., Lee, B.H., 2014. Enhanced conidial thermotolerance and field efficacy of *Beauveria bassiana* by saccharides. Biotechnol. Lett. 36 (12), 2481–2488.

Lopes, R.B., Alves, S.B., Tamai, M.A., 2000. Control of *Frankliniella occidentalis* in hidroponic lettuce by *Metarhizium anisopliae*. Sci. Agric. 57 (2), 239–243.

Loughner, R.L., Warnock, D.F., Cloyd, R.A., 2005. Resistance of greenhouse, laboratory, and native populations of western flower thrips to spinosad. Hortscience 40 (1), 146–149.

Macintyre Allen, J.K., Scott-Dupree, C.D., Tolman, J.H., Ron Harris, C., 2005. Resistance of *Thrips tabaci* to pyrethroid and organophosphorus insecticides in Ontario. Pest Manag. Sci. 61, 809–815.

Macleod, D.M., Tyrrell, D., Carl, K.P., 1976. Entomophthora parvispora sp. nov., a pathogen of *Thrips tabaci*. Entomophaga 21, 307–312.

Maniania, N.K., Ekesi, S., Loehr, B., Mwangi, F., 2002. Prospects for biological control of the western flower thrips, *Frankliniella occidentalis*, with the entomopathogenic fungus, *Metarhizium anisopliae*, on chrysanthemum. Mycopathologia 155, 229–235.

Maniania, N.K., Sithanantham, S., Ekesi, S., Ampong-Nyarko, K., Baumgärtner, J., Löhr, B., Matoka, C.M., 2003. A field trial of the entomopathogenous fungus *Metarhizium anisopliae* for control of onion thrips, *Thrips tabaci*. Crop Prot. 22, 553–559.

Martin, N.A., Workman, P.J., Butler, R.C., 2003. Insecticide resistance in onion thrips (*Thrips tabaci*) (Thysanoptera: Thripidae). N. Z. J. Crop. Hortic. Sci. 31, 99–106.

Messelink, G., van Steenpaal, S., van Wensveen, W., 2005. *Typhlodromips swirskii* (Athias-Henriot) (Acari: Phytoseiidae): a new predator for thrips control in greenhouse cucumber. IOBC/WPRS Bull. 28 (1), 183–186.

Morishita, M., 2008. Pyrethroid-resistant onion thrips, *Thrips tabaci* Lindeman (Thysanoptera: Thripidae), infesting persimmon fruit. Appl. Entomol. Zool. 43, 25–31.

Morse, J.G., Hoddle, M.S., 2006. Invasion biology of thrips. Annu. Rev. Entomol. 51 (1), 67–89.

Mouden, S., Sarmiento, K.F., Klinkhamer, P.G., Leiss, K.A., 2017. Integrated pest management in western flower thrips: past, present and future. Pest Manag. Sci. 73 (5), 813–822.

Murai, T., 2001. Development and reproductive capacity of *Thrips hawaiiensis* (Thysanoptera: Thripidae) and its potential as a major pest. Bull. Entomol. Res. 91, 193–198.

Murai, T., Loomans, A.J.M., 2001. Evaluation of an improved method for mass-rearing of thrips and a thrips parasitoid. Entomol. Exp. Appl. 101, 281–289.

Murphy, B.C., Morisaw, T.A., Newman, J.P., Tjosvold, S.A., Parrella, M.P., 1998. Fungal pathogen controls thrips in greenhouse flowers. Calif. Agric. 52 (3), 32–36.

Nabirye, J., Nampala, P., Kyamanywa, S., Ogenga-Latigo, M.W., Wilson, H., Adipala, E., 2003. Determination of damage-yield loss relationships and economic injury levels of flower thrips on cowpea in eastern Uganda. Crop Prot. 22, 911–915.

Naik, P.R.H., Shekharappa, 2009. Field evaluation of different entomopathogenic fungal formulations against sucking pests of okra. Karnataka J. Agric. Sci. 22, 575–578.

Ngakou, A., Tamo, M., Parh, I.A., Nwagad, D., Ntonifore, N.N., Korieb, S., Nebanee, C.L.N., 2008. Management of cowpea flower thrips, *Megalurothrips sjostedti* (Thysanoptera, Thripidae), in Cameroon. Crop Prot. 27, 481–488.

Niassy, S., Maniania, N.K., Subramanian, S., Gitonga, L.M., Ekesi, S., 2012. Performance of a semiochemical-baited autoinoculation device treated with *Metarhizium anisopliae* for

control of *Frankliniella occidentalis* on French bean in field cages. Entomol. Exp. Appl. 142, 97–103.
North, J.P., Cuthbertson, A.G.S., Walters, K.F.A., 2006. The effcacy of two entomopathogenic biocontrol agents against adult *Thrips palmi* (Thysanoptera: Thripidae). J. Invertebr. Pathol. 92, 89–92.
Northfield, T.D., Paini, D.R., Funderburk, J.O., Reitz, S.R., 2008. Annual cycles of *Frankliniella* spp. (Thysanoptera: Thripidae) thrips abundance on North Florida uncultivated reproductive hosts: predicting possible sources of pest outbreaks. Ann. Entomol. Soc. Am. 101 (4), 769–778.
Ong, T.W.Y., Vandermeer, J.H., 2015. Coupling unstable agents in biological control. Nat. Commun. 6.
Ortiz-Urquiza, A., Keyhani, N.O., 2013. Action on the surface: entomopathogenic fungi versus the insect cuticle. Insects 4 (3), 357–374.
Pizzol, J., Nammour, D., Hervouet, P., Bout, A., Desneux, N., Mailleret, L., 2010. Comparison of two methods of monitoring thrips populations in a greenhouse rose crop. J. Pestic. Sci. 83, 191–196.
Rangel, D.E., Fernandes, E.K., Dettenmaier, S.J., Roberts, D.W., 2010. Thermotolerance of germlings and mycelium of the insect-pathogenic fungus *Metarhizium* spp. and mycelial recovery after heat stress. J. Basic Microbiol. 50 (4), 344–350.
Reitz, S.R., 2009. Biology and ecology of the western flower thrips (Thysanoptera: Thripidae): the making of a pest. Fla. Entomol. 92 (1), 7–13.
Reitz, S.R., Yearby, E.L., Funderburk, J.E., Stavisky, J., Timur Momol, M., Olson, S.M., 2003. Integrated management tactics for *Frankliniella* thrips (Thysanoptera: Thripidae) in field-grown pepper. J. Econ. Entomol. 96 (4), 1201–1214.
Riley Jr., D., Sparks, A., Srinivasan, R., Kennedy, G., Fonsah, G., Scott, J., Olson, S., 2018. Chapter 3—Thrips: biology, ecology, and management. In: Sustainable Management of Arthropod Pests of Tomato. Academic Press, Salt Lake City, UT, pp. 49–71.
Riley, D.G., Joseph, S.V., Srinivasan, R., Diffie, S., 2011. Thrips vectors of tospoviruses. J. Integr. Pest Manag. 1 (2), 1–10.
Riudavets, J., 1995. Predators of *Frankliniella occidentalis* (Perg.) and *Thrips tabaci* Lind.: a review. Wageningen Agric. Univ. Papers 31 (4), 577–579.
Romero-González, R.R., Mirnezhad, M., Leiss, K.A., Choi, Y.H., Verpoorte, R., Klinkhamer, P.G.L., 2010. Metabolomic analysis of host–plant resistance to thrips in wild and domesticated tomatoes. Phytochem. Anal. 21, 110–117.
Rosenheim, J.A., Welter, S.C., Johnson, M.W., Mau, R.F.L., Gusukuma Minuto, L.R., 1990. Direct feeding damage on cucumber by mixed-species infestations of *Thrips palmi* and *Frankliniella occidentalis* (Thysanoptera: Thripidae). J. Econ. Entomol. 84 (3), 1519–1525.
Rotenberg, D., Jacobson, A.L., Schneweis, D.J., Whitfield, A.E., 2015. Thrips transmission of tospoviruses. Curr. Opin. Virol. 15, 80–89.
Roy, H.E., Pell, J.K., 2000. Interactions between entomopathogenic fungi and other natural enemies: implications for biological control. Biocontrol Sci. Technol. 10, 737–752.
Saito, T., 1991. A field trial of an entomopathogenic fungus, *Beauveria bassiana* (Balsamo) Vuillemin., for the control of *Thrips palmi* Karny (Thysanoptera, Thripidae). Jpn J. Appl. Entomol. Zool. 35, 80–81.
Saito, T., 1992. Control of *Thrips palmi* and *Bemisia tabaci* by a mycoinsecticidial preparation of *Verticillium lecanii*. Proc. Kanto-Tosan Plant Protect Soc. 39, 209–210.
Saito, T., Kubota, S., Shimazu, M., 1989. A first record of Entomophathogenic fungus, *Neozygites parvispora* (Macleod & Carl) Rem. & Kell., on *Thrips palmi* Karny (Thysanoptera: Thripidae) in Japan. Appl. Entomol. Zool. 24 (2), 233–235.
Samson, R.A., Ramakers, P.M.J., Oswaid, T., 1979. *Entomophthora thripidum*, a new fungal pathogen of *Thrips tabaci*. Can. J. Bot. 57, I317–1323.

Sánchez-Peña, S.R., San-Juan Lara, J., Medina, R.F., 2011. Occurrence of entomopathogenic fungi from agricultural and natural ecosystems in Saltillo, México, and their virulence towards thrips and whiteflies. J. Insect Sci. 11, 1–10.

Seal, D.R., Klassen, W., Sabines, C., 2007. Efficacy of BotaniGard®, TriCon™ and *Metarhizium anisopliae* treatments for the control of chilli thrips, *Scirtothrips dorsalis* hood (Thysanoptera: Thripidae) in the greenhouse. Proc. Caribbean Food Crops Soc. 43, 30–38.

Seiedy, M., Tork, M., Deyhim, F., 2015. Effect of the entomopathogenic fungus *Beauveria bassiana* on the predatory mite *Amblyseius swirskii* (Acari: Phytoseiidae) as a non-target organism. Syst. Appl. Acarol. 20 (3), 241–250.

Sengonca, C., Thungrabeab, M., Blaeser, P., 2006. Potential of different isolates of entomopathogenic fungi from Thailand as biological control agents against western flower thrips, *Frankliniella occidentalis* (Pergande) (Thysanoptera: Thripidae). J. Plant Dis. Protect. 113 (2), 74–80.

Shah, P.A., Pell, J.K., 2003. Entomopathogenic fungi as biological control agents. Appl. Microbiol. Biotechnol. 61, 413–423.

Shelton, A.M., Zhao, J.Z., Nault, B.A., Plate, J., Musser, F.R., Larentzaki, E., 2006. Patterns of insecticide resistance in onion thrips (Thysanoptera: Thripidae) in onion Þelds in New York. J. Econ. Entomol. 99, 1798–1804.

Shipp, J.L., Hao, X., Papadopoulos, A.P., Binns, M.R., 1998. Impact of western flower thrips (Thysanoptera: Thripidae) on growth, photosynthesis and productivity of greenhouse sweet pepper. Sci. Hortic. 72, 87–102.

Shipp, L., Kapongo, J.P., Park, H.H., Kevan, P., 2012. Effect of bee-vectored *Beauveria bassiana* on greenhouse beneficials under greenhouse cage conditions. Biol. Control 63, 135–142.

Silva, A.L.E., Murguido Morales, C.A., Torres, M.M., 2011. Patogenicidad de los hongos *Metarhizium anisopliae* (metschn.), *Lecanicillium lecanii* (zimm.) zare & gams y *Beauveria bassiana* (bals.-criv.) vuill. sobre *Thrips palmi* karny en el cultivo de la papa (solanum tuberosum l.). Fitosanidad 15 (3), 147–151.

Silvia Yusuf, E., Nuryani, W., Djatnika, I., 2010. The effect of several carriers on *Beauveria bassiana* to control Thrips parvispinus Karny on chrysanthemum under plastichouse. J. Hort. 20 (1), 80–85.

Skinner, M., Parker, B.L., Bergdahl, D.R., 1991. *Verticillium lecanii*, isolated from larvae of pear thrips, *Taeniothrips inconsequens*, in Vermont. J. Invertebr. Pathol. 58, 157–163.

Skinner, M., Gouli, S., Frank, C.E., Parker, B.L., Kim, J.S., 2012. Management of *Frankliniella occidentalis* (Thysanoptera: Thripidae) with granular formulations of entomopathogenic fungi. Biol. Control 63, 246–252.

Stuart, R.R., Gao, Y.L., Lei, Z.R., 2011. Thrips: pests of concern to China and the United States. Agric. Sci. China 10 (6), 867–892.

Tamò, M., Baumgärtner, J., Delucchi, V., Herren, H.R., 1993. Assessment of key factors responsible for the pest status of the bean flower thrips *Megalurothrips sjostedti* (Thysanoptera: Thripidae) in West Africa. Bull. Entomol. Res. 83 (2), 251–258.

Tang, L.D., Yan, K.L., Fu, B.L., Wu, J.H., Liu, K., Lu, Y.Y., 2015. The life table parameters of *Megalurothrips usitatus* (Thysanoptera: Thripidae) on four leguminous crops. Fla. Entomol. 98 (2), 620–625.

Tang, L.D., Zhao, H.Y., Fu, B.L., Han, Y., Liu, K., Wu, J.H., 2016. Colored sticky traps to selectively survey thrips (Thysanoptera) in cowpea ecosystem. Neotrop. Entomol. 45 (1), 96–101.

Thungrabeab, M., Tongma, S., 2007. Effect of entomopathogenic fungi, *Beauveria bassiana* (Balsam) and *Metarhizium anisopliae* (Metsch) on non target insects. KMITL Sci. Technol. J. 7, 8–12.

Thungrabeab, M., Blaeser, P., Sengonca, C., 2006. Possibilities for biocontrol of the onion thrips *Thrips tabaci* Lindeman (Thys., Thripidae) using different entomopathogenic fungi from Thailand. Mitt. Dtsch. Ges. Allg. Angew. Ent. 15, 299–304.

Toledo-Hernández, R.A., Ortiz-Girón, J.A., Sánchez, D., 2017. Preliminary observations on pathogenicity of commercial formulations of *Metarhizium anisopliae* and *Beauveria bassiana* entomopathogenic fungi for control of *Frankliniella invasor* under laboratory conditions. Southwest. Entomol 42 (4), 1035–1040.

Ugine, T.A., Wraight, S.P., Brownbridge, M., Sanderson, J.P., 2005. Development of a novel bioassay for estimation of median lethal concentrations (LC_{50}) and doses (LD_{50}) of the entomopathogenic fungus *Beauveria bassiana*, against western flower thrips, *Frankliniella occidentalis*. J. Invertebr. Pathol. 89, 210–218.

Ugine, T.A., Wraight, S.P., Sanderson, J.P., 2006. Influences of impatiens pollen and exposure to *Beauveria bassiana* on bionomics of western flower thrips *Frankliniella occidentalis*. Biol. Control 37, 186–195.

Ugine, T.A., Wraight, S.P., Sanderson, J.P., 2007. Effects of manipulating spray-application parameters on efficacy of the entomopathogenic fungus *Beauveria bassiana* against western flower thrips, *Frankliniella occidentalis*, infesting greenhouse impatiens crops. Biocontrol Sci. Technnol. 17, 193–219.

Vacante, V., Cacciola, S.O., Pennisi, A.M., 1994. Epizootiological study of *Neozygites parvispora* (Zygomycota: Entomophthoraceae) in a population of *Frankliniella occidentalis* (Thysanoptera, Thripidae) on pepper in sicily. Entomophaga 39, 123–130.

Van der Schaaf, E.D., Malais, M., Ravensberg, W., 1990. In: The use of *Verticillium lecanii* against whitefly and thrips in glasshouse vegetables in the Netherlands. Proceedings and Abstracts, 5th International Colloquium on Invertebrate Pathology and Microbial Control. Adelaide, Australia, Glen Osmond, Australia.

van Lenteren, J.C., Roskam, M.M., Timmer, R., 1997. Commercial mass production and pricing of organisms for biological control of pests in Europe. Biol. Control 10, 143–149.

Vestergaard, S., Gillespie, A.T., Butt, T.M., Chreiter, G., Eilenberg, J., 1995. Pathogenicity of hyphomycete fungi *Verticillium lecanii* and *Metarhizium anisopliae* to the western flower thrips, *Frankliniella occidentalis*. Biocontrol Sci. Technol. 5, 185–192.

Wang, Y.H., Zheng, C.Y., Wang, J.P., 2011a. Virulence of *Beauveria bassiana* to *Frankliniella occidentalis* adults and scanning electron microscopic observation on its infection process. Chin. J. Biol. Control. 27 (3), 324–330.

Wang, Z.H., Hou, W.J., Hao, C.Y., Wu, Q.J., Xu, B.Y., Zhang, Y.J., 2011b. Monitoring the insecticide resistance of the field populations of western flower thrips, *Frankliniella occidentalis* in Beijing area. Chin. J. Appl. Entomol. 48 (3), 542–547.

Wu, S.Y., Gao, Y.L., Xu, X.N., Zhang, Y.P., Wang, J., Lei, Z.R., Smagghe, G., 2013. Laboratory and greenhouse evaluation of a new entomopathogenic strain of *Beauveria bassiana* for control of the onion thrips *Thrips tabaci*. Biocontrol Sci. Technnol. 23 (7), 794–802.

Wu, S.Y., Gao, Y.L., Zhang, Y.P., Wang, E.D., Xu, X.N., Lei, Z.R., 2014. An entomopathogenic strain of *Beauveria bassiana* against *Frankliniella occidentalis* with no detrimental effect on the predatory mite *Neoseiulus barkeri*: evidence from laboratory bioassay and scanning electron microscopic observation. PLoS One 9 (1), e84732.

Wu, S.Y., Gao, Y.L., Xu, X.N., Goettel, M.S., Lei, Z.R., 2015. Compatibility of *Beauveria bassiana* with *Neoseiulus barkeri* for control of *Frankliniella occidentalis*. J. Integr. Agric. 14 (1), 98–105.

Wu, S.Y., Gao, Y.L., Smagghe, G., Xu, X.N., Lei, Z.R., 2016. Interactions between the entomopathogenic fungus *Beauveria bassiana* and the predatory mite *Neoseiulus barkeri* and biological control of their shared prey/host *Frankliniella occidentalis*. Biol. Control 98, 43–51.

Wu, S.Y., He, Z., Wang, E.D., Xu, X.N., Lei, Z.R., 2017. Application of *Beauveria bassiana* and *Neoseiulus barkeri* for improved control of *Frankliniella occidentalis* in greenhouse cucumber. Crop Prot. 96, 83–87.

Ying, S.H., Feng, M.G., 2006. Medium components and culture conditions affect the thermotolerance of aerial conidia of fungal biocontrol agent *Beauveria bassiana*. Lett. Appl. Microbiol. 43 (3), 331–335.

Zahn, D.K., Haviland, D.R., Stanghellini, M.E., Morse, J.G., 2013. Evaluation of *Beauveria bassiana* for Management of Citrus Thrips (Thysanoptera: Thripidae) in California blueberries. J. Econ. Entomol. 106 (5), 1986–1995.

Zhang, T., Reitz, S.R., Wang, H., Lei, Z.R., 2015. Sublethal effects of *Beauveria bassiana* (Ascomycota: Hypocreales) on life table parameters of *Frankliniella occidentalis* (Thysanoptera: Thripidae). J. Econ. Entomol. 108 (3), 975–985.

Zhao, G.Y., Liu, W., Brown, J.M., Knowles, C.O., 1995. Insecticide resistance in field and laboratory strains of western flower thrips (Thysanoptera: Thripidae). J. Econ. Entomol. 88, 1164–1170.

MIX
Papier | Fördert
gute Waldnutzung
FSC® C083411

Kolibri 360 GmbH
Ferdinand-Jühlke-Straße 7
99095 Erfurt, Deutschland
produktsicherheit@kolibri360.de